Study Guide for

INTRODUCTION TO Chemistry

and

INTRODUCTION TO Chemistry

Extended Edition

Hardwick • Bouillon

Joseph Ledbetter

Contra Costa College

SAUNDERS COLLEGE PUBLISHING

Harcourt Brace College Publishers

Fort Worth Philadelphia San Diego New York Orlando Austin
San Antonio Toronto Montreal London Sydney Tokyo

Printed in the United States of America.

Ledbetter: Study Guide to accompany INTRODUCTION TO CHEMISTRY and INTRODUCTION TO CHEMISTRY, EXTENDED EDITION, by Hardwick and Bouillon

ISBN 0-03-029053-8

345 021 987654321

Acknowledgements

I thank Sandi Kiselica of Saunders College Publishing for managing the publication of this study guide. I also thank Michelle Floyd, Adeliza Flores and Dr. Steven Albrecht for a careful reading of the manuscript. My highest praise goes to Kim A. Meyer for her proofreading ability, aesthetic sensibilities, superb graphics skills and encouraging support during the stressful times.

For The Student

- This Study Guide accompanies the text *Introduction to Chemistry* and *Introduction to Chemistry, Extended Edition* by Hardwick and Bouillon. For each chapter in the text, this guide has the following sections:

  A. Outline and Study Hints
  B. Solutions to In-Text Exercises
  C. Practice Test
  D. Answers to the Practice Test

- The **"Outline and Study Hints"** is a concise summary of each section as well as a listing of recommended hints such as what should be memorized.

- The **"Solutions to In- Text Exercises"** are detailed solutions to the exercises that are within the text. These are to be consulted only after attempting the exercise by jotting down your thoughts and conclusions. Do not simply follow the solutions without first writing down some of your own ideas.

- The **"Practice Test"** should be attempted after understanding the examples and exercises in the text. Answers are included to check your comprehension of the text material and applications to the exercises.

Table of Contents

Chapter 1

Measurement and Problem Solving in Chemistry

A. Outline and Study Hints

Section 1.1

- Chemistry is concerned with the quantitative measurement of the properties of matter.
- A measured value must include units. Every measurement has some degree of error or uncertainty.

Section 1.2

- Uncertainty is indicated by reporting a measurement to the correct number of significant figures.
- The number of significant figures in a measurement is the number of digits that are certain plus one more "estimated" digit.
- The rules 1-6 for writing and interpreting significant figures on page 11of text should be memorized.
- When performing calculations, the final result should be "rounded" to no more significant figures than is justified by the measurement with the fewest significant figures (for multiplications and divisions) or the fewest decimal places (for addition and subtraction).

Section 1.3

- Scientific notation is most useful when expressing measurements that have very large or small values.
- To convert a decimal value to scientific notation, move the decimal point to the right or left to yield a value that is between 1 and 10. The exponent is the number of places that you moved the decimal.
- A negative exponent is used if you moved the decimal point to the right in the above process. For example $0.0045 = 4.5 \times 10^{-3}$. A positive exponent is used if you moved the decimal point to the left, $45000 = 4.5 \times 10^{4}$.
- Conversion of scientific notation values to decimal values uses the reverse of the above procedure.
- All the digits in a measurement reported in scientific notation are significant. For example 8.0×10^{6} has two significant figures whereas 8×10^{6} has only one significant figure.

Section 1.4

- The organized six step approach to problem solving is helpful and should be followed.
- Conversion factors, also called unit factors, are quotients whose numerators and denominators are equal. An example is the statement that there are 16 ounces in a pound. This can be written as a ratio

$$\frac{16 \text{ ounces}}{\text{pound}} \text{ or } \frac{\text{pound}}{16 \text{ ounces}}$$

- A "solution map" for a problem is a step-by-step program that organizes your problem solving approach. It is important that you write these steps out. This will help you to correct errors. For example, if your answer to a problem states that a man weighs 150 pounds that is reasonable. But if your answer claims the man to be 1500 pounds, you know that a calculation was done incorrectly.
- Estimations are often helpful when you analyze your answer for reasonableness.

Section 1.5

- Conversion factors are used in solving many kinds of problems.
- Many problems can be solved by applying a single conversion factor. For example, converting two and a half tons to pounds requires the conversion factor that states that two thousand pounds is equivalent to one ton

$$\frac{\text{ton}}{2000\text{ lbs}} \text{ or } \frac{2000\text{ lbs}}{\text{ton}}$$

The solution map is

$$\boxed{\text{tons}} \rightarrow \boxed{\text{pounds}}$$

and the multiplication showing the cancellation of the units is

$$2.5\ \cancel{\text{tons}} \times \frac{2000\text{ lbs}}{\cancel{\text{ton}}} = 5000\ \text{lbs}$$

Therefore, 2.5 tons equals 5.0×10^3 lbs. The answer 5000 lbs does not clearly state the correct significant figures. But note that 5.0×10^3 has two significant figures due to 2.5 tons.

- Units are handled algebraically like simple variables such as *x* or *y*. For example, multiplying density (pounds per cubic feet, lbs/ft^3) and area (square feet, ft^2) yields pounds per foot:

$$\frac{\text{lbs}}{\text{ft}^3} \times \text{ft}^2 = \text{lbs} \times \frac{\text{ft}^2}{\text{ft}^3} = \text{lbs} \times \frac{1}{\text{ft}} = \frac{\text{lbs}}{\text{ft}}$$

Here's another example of a problem using conversion factors. Converting from six square feet into square inches requires that the conversion factor be squared,

$$\left(\frac{12\text{ in}}{\text{ft}}\right)^2 = \frac{12^2\text{ in}^2}{1\text{ ft}^2} = \frac{144\text{ in}^2}{\text{ft}^2}$$

Thus

$$6\,00\ \cancel{\text{ft}^2} \times \frac{144\text{ in}^2}{\cancel{\text{ft}^2}} = 864\text{ in}^2$$

Section 1.6

- The metric system of units are used in most sciences including chemistry. The fundamental units and their abbreviations given in Table 1.1 of the text should be memorized. The prefixes in Table 1.2 should also be committed to memory.

Section 1.7

- Measurements of length use the meter and prefixes. Some conversion factors such as

$$\frac{2.54\text{ cm}}{\text{in}}$$

are important enough to memorize. This can be used to convert 10.0 miles per hour to meters per second. The solution map is

$$\boxed{\frac{\text{miles}}{\text{hour}}} \rightarrow \boxed{\frac{\text{feet}}{\text{hour}}} \rightarrow \boxed{\frac{\text{inches}}{\text{hour}}} \rightarrow \boxed{\frac{\text{centimeters}}{\text{hour}}} \rightarrow \boxed{\frac{\text{meters}}{\text{hour}}} \rightarrow \boxed{\frac{\text{meters}}{\text{seconds}}}$$

$$\frac{10.0\text{ mi}}{\text{hr}} \times \frac{5280\text{ ft}}{\text{mi}} \times \frac{12\text{ in}}{\text{ft}} \times \frac{2.54\text{ cm}}{\text{in}} \times \frac{\text{m}}{100\text{ cm}} \times \frac{\text{hr}}{3600\text{ s}} = 4.47\frac{\text{m}}{\text{s}}$$

The final answer has been rounded off to three significant figures.

Section 1.8

- Volume units are cubic length units such as cm^3; however, a defined volume unit, the liter (L), is more commonly used in chemistry. The conversion factor

$$\frac{\text{L}}{1000\text{ cm}^3} \text{ or } \frac{\text{L}}{1000\text{ mL}}$$

should be memorized. Remember that mL symbolizes milliliters.

Section 1.9

- Although the SI metric unit for mass is the kilogram, chemists more often use grams. One thousand grams equals one kilogram.

$$\frac{1000\text{ g}}{\text{kg}} \text{ or } \frac{\text{kg}}{1000\text{ g}}$$

Section 1.10

- Three temperature scales are commonly used in chemistry: Celsius, °C, Kelvin, K and Fahrenheit, °F.
- The Kelvin scale is related to the Celsius scale by

$$\text{K} = {}^\circ\text{C} + 273.$$

In addition, the Fahrenheit and Celsius scales are related by

$${}^\circ\text{F} = 32 + \frac{9}{5}{}^\circ\text{C} \text{ and } {}^\circ\text{C} = \frac{5}{9} \times ({}^\circ\text{F}-32).$$

Section 1.11

- Density is a property of a material that depends on temperature. The units of density are mass divided by volume such as

$$\frac{\text{g}}{\text{cm}^3} \text{ grams per cubic centimeter OR } \frac{\text{g}}{\text{L}} \text{ grams per liter}$$

B. Solutions to In-Text Exercises

Exercise 1.1

a) $4.2 \times 0.311 = 1.3\cancel{062} = 1.3$; rounded off to two significant figures due to the 4.2 value.

b) $\frac{74}{8} = 9.25$ provided that both 74 and 8 are exact or $\frac{74}{8} = 9$ if rounded off to one significant figure.

c) $\frac{4.42 \times 0.8843 \times 19}{36257} = 0.0020$ (two significant figures because of 19).

d) $0.81 \times 788.3 \times 0.000344 = 0.22$ (two significant figures because of 0.81).

Exercise 1.2

(a)	(b)	(c)	(d)	(e)
33.482				
0.88	553.1			3432
2.9993	0.22	51.334	344.28544	240
+410.33	+25.334	-7.8832	-91.	21.7
447.69	578.7	43.451	253	3690
	(rounded up)	(rounded up)		(rounded down to 3 significant figures)

Exercise 1.3

$(2.3 + 4.00) \times (3.6 + 9) = 79$ (2 significant figures).

Exercise 1.4

a. $7.04 \times 10^{-4} = 0.000704$

b. $5.20 \times 10^{-2} = 0.0520$

c. $8.03 \times 10^{5} = 803{,}000$

d. $1.003 \times 10^{2} = 100.3$

Exercise 1.5

The solution map is

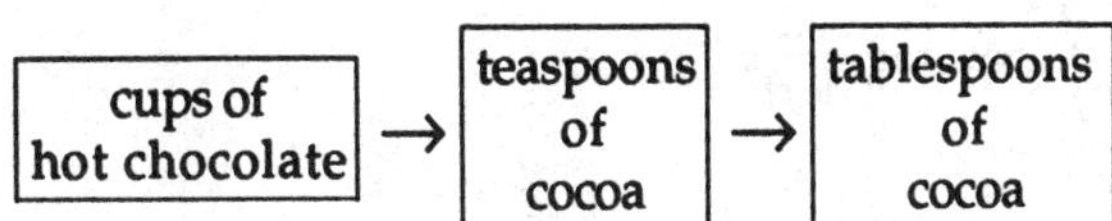

$$5 \text{ cups} \times \frac{4 \text{ teaspoons}}{\text{cup}} \times \frac{\text{tablespoon}}{3 \text{ teaspoons}} = 7 \text{ tablespoons}$$ {rounded off to one significant figure}

Exercise 1.6

The solution map is miles → feet → yards

$$0.242 \text{ mi} \times \frac{5280 \text{ ft}}{\text{mi}} \times \frac{\text{yd}}{3 \text{ ft}} = 426 \text{ yd}$$

Note that 3 ft is not considered in the determination of significant figures because it defines a conversion factor, not a measurement such as 0.242 mi.

Exercise 1.7

The solution map

yd^3 → ft^3 → in^3 → gal

first requires us to compute the volume of a rectangular shaped container of water. The volume is the width times the length times the depth:

$$1.32 \text{ yd} \times 3.55 \text{ yd} \times 0.23 \text{ yd} = 1.08 \text{ yd}^3$$

We have carried one extra significant figure because we are not finished with our calculations. Thus,

$$1.08 \text{ yd}^3 \times \left(\frac{3 \text{ ft}}{\text{yd}}\right)^3 \times \left(\frac{12 \text{ in}}{\text{ft}}\right)^3 \times \frac{\text{gal}}{231 \text{ in}^3} = 220 \text{ gal}$$

The final answer has two significant figures because our original value of depth was 0.23 yd.

Exercise 1.8

After calculating the area of the wall $10.5 \text{ ft} \times 8.0 \text{ ft} = 84 \text{ ft}^2$, we apply solution map

ft^2 → yd^2 → rolls

$$84 \text{ ft}^2 \times \left(\frac{\text{yd}}{3 \text{ ft}}\right)^2 \times \frac{\text{roll}}{4.0 \text{ yd}^2} = 2.3 \text{ or } 3 \text{ rolls}$$

Remember to square the 3 in the expression $\left(\frac{\text{yd}}{3 \text{ ft}}\right)^2$ yielding $\frac{\text{yd}^2}{9 \text{ ft}^2}$.

Exercise 1.9

Our solution map

ft^3 → in^3 → oz → lbs

becomes

$$1.00 \text{ ft}^3 \times \left(\frac{12 \text{ in}}{\text{ft}}\right)^3 \times \frac{4.55 \text{ oz}}{\text{in}^3} \times \frac{\text{lb}}{16 \text{ oz}} = 491 \text{ lbs}$$

Exercise 1.10

Notice that the question asks for an answer in square feet, yet the volume is given in cubic yards and the thickness is given in inches. We thus first convert all quantities to feet units:

$\boxed{yd^3} \rightarrow \boxed{ft^3}$

$$\text{volume} = 0.050\ yd^3 \times \left(\frac{3\ ft}{yd}\right)^3 = 1.35\ ft^3$$

$\boxed{in} \rightarrow \boxed{ft}$

$$\text{thickness} = 0.25\ in \times \frac{ft}{12\ in} = 2.08 \times 10^{-2}\ ft \text{ (we will carry one extra digit).}$$

Now dividing the volume by the thickness will yield the area of the topsoil

$$\frac{\boxed{\text{volume } (ft^3)}}{\boxed{\text{thickness (ft)}}} = \boxed{\text{area } (ft^2)} \text{ notice that the units are in square footage.}$$

$$\frac{1.35\ ft^3}{2.08 \times 10^{-2}\ ft} = 65\ ft^2\text{, which is reasonable.}$$

Exercise 1.11

The solution map is

$\boxed{Å} \rightarrow \boxed{m} \rightarrow \boxed{cm}$

$$1\ Å \times \frac{m}{10^{10}\ Å} \times \frac{100\ cm}{m} = 1 \times 10^{-8}\ cm$$

Exercise 1.12

The solution map is

$\boxed{mm} \rightarrow \boxed{cm} \rightarrow \boxed{in} \rightarrow \boxed{ft}$

$$290\ mm \times \frac{cm}{10\ mm} \times \frac{in}{2.54\ cm} \times \frac{ft}{12\ in} = 0.95\ ft\text{, which is less than one foot, so the glass will fit.}$$

Exercise 1.13

The solution map converting to meters is

$\boxed{mi} \rightarrow \boxed{ft} \rightarrow \boxed{in} \rightarrow \boxed{cm} \rightarrow \boxed{m}$ then $\rightarrow \boxed{km}$

$$25{,}000\ mi \times \frac{5280\ ft}{mi} \times \frac{12\ in}{ft} \times \frac{2.54\ cm}{in} \times \frac{m}{100\ cm} = 4.0 \times 10^7\ m$$

$$(4.0 \times 10^7\ m) \times \frac{km}{1000\ m} = 4.0 \times 10^4\ km$$

Exercise 1.14

In order to answer this question you need to calculate how many liters are contained in $0.75\ m^3$. The solution map is (using $1\ m^3 = 10^3\ L$).

$$\boxed{m^3} \rightarrow \boxed{L}$$

thus,

$0.75\ m^3 \times \frac{10^3\ L}{m^3} = 750\ L$ but on each trip you carry 2 .0 L in a jug.

Thus $750\ L \times \frac{1\ trip}{2.0\ L} = 375$ or 380 trips rounded up to the correct significant figures.

Exercise 1.15

The solution map becomes

$$\boxed{oz} \rightarrow \boxed{qt} \rightarrow \boxed{L} \rightarrow \boxed{dL} \rightarrow \boxed{Lire}$$

$$12.0\ oz \times \frac{qt}{32\ oz} \times \frac{0.946\ L}{qt} \times \frac{10\ dL}{L} \times \frac{1400\ Lire}{5.0\ dL} = 990\ Lire$$

Exercise 1.16

Performing the calculation requires a conversion of 0.5 lbs into hundreds of grams, thus the solution map is included in the calculation.

$0.5\ lbs \times \frac{453.6\ g}{lb} = 227\ g$ or you should buy two hundred grams.

Exercise 1.17

$$76\ kg \times \frac{1000\ g}{kg} = 76{,}000\ g \text{ or } 7.6 \times 10^4\ g$$

Exercise 1.18

$K = {}^\circ C + 273$ is the relationship allowing us to calculate the Kelvin equivalent of Celsius temperatures.

$K = 32 + 273 = 305\ K$

$K = 178 + 273 = 451\ K$

Exercise 1.19

${}^\circ C = K - 273$ or

${}^\circ C = 212 - 273 = -61\ {}^\circ C$

${}^\circ C = 298 - 273 = 25\ {}^\circ C$ which is about room temperature.

Exercise 1.20

Remember that density is the ratio of the mass of a sample to its volume.

$\frac{32.4\ g}{21.0\ mL} = 1.54\ g/mL$ is the density of the unknown substance.

Exercise 1.21

The density is the mass per unit volume

$$d_{H_2O}^{50\ °C} = \frac{32.7\ g}{33.1\ mL} = 0.988\ \frac{g}{mL}$$

which is less than the density of water at 4 °C (1.00 g/mL).

Exercise 1.22

The volume (V) of a rectangular shaped solid is the length (l) times the width (w) times the thickness (t).

$$(1)\ \ V = l \times w \times t$$

The volume is also the mass (m) divided by the density (d).

$$(2)\ \ V = \frac{m}{d}$$

Equating (1) and (2) yields

$$l \times w \times t = \frac{m}{d}.$$

Solving for t by dividing by $l \times w$ yields

$$t = \frac{m}{d \times l \times w}.$$

Inserting the quantities yields the thickness

$$t = \frac{12.74\ g}{8.92\ \frac{g}{cm^3} \times 7.8\ cm \times 3.5\ cm} = 0.052\ cm.$$

Exercise 1.23

The conversion uses the density to compute the volume of a given mass after converting to grams. Therefore the solution map is

$$\boxed{kg} \rightarrow \boxed{g} \rightarrow \boxed{cm^3}$$

$$3.31\ kg \times \frac{1000 g}{kg} \times \frac{cm^3}{2.16\ g} = 1.53 \times 10^3\ cm^3 \text{ or } 1.53\ L.$$

C. Chapter 1 Practice Test

1. Perform the following calculations and report the answers to the correct number of significant figures and use scientific notation if appropriate.
   a. $6.02 \times 4.9 =$
   b. $68.24 - 29.1 =$
   c. $\dfrac{(6.023 \times 10^{23}) \times (1.619 \times 10^{-19})}{1.8 \times 10^{-3}} =$

2. List the five fundamental SI units and at least five metric prefixes.

3. Perform the following conversions after drawing a solution map for the problem
   a. Calculate the length in inches of a 100.0 mm ruler.
   b. Calculate your mass in kilograms.
   c. Calculate the diameter in nanometers of a sphere 10.0 Å in diameter.
   d. Calculate the area in square centimeters of an $8\frac{1}{2} \times 11$ inch sheet of paper.

4. Liquid nitrogen boils at about 77 K. Convert this temperature to °C and °F.

5. Calculate the mass of twenty gallons of gasoline which has a density of about $0.7 \frac{g}{mL}$

6. Which of the following is equivalent to 1 cm^3?
   a. 1 mL b. 1 mm c. 1 mg d. 1 g

7. Which of the following is the density of kerosene if a 10.0 g sample occupies 13.3 mL?
   a. 0.075 g/mL b. 0.752 g/mL c. 1.33 g/mL d. 13.3 g/mL

8. Which of the following is a length unit?
   a. mL b. mg c. dm d. cc

9. A temperature in Kelvins is always a (smaller/larger) number than the equivalent Celsius temperature.

10. Chemists often use __________ as a rough measure of the uncertainty of a quantity.

D. Answers To Practice Test:

1. a. 29 b. 39.1 c. 5.4×10^7

2. See Tables 1.1 and 1.2.

3. You'll need to use the text for most of the conversion factors. Hint, 1 cm = 10 mm.

   a. [mm] → [cm] → [inches]

$$100.0\ \text{mm} \times \frac{\text{cm}}{10\ \text{mm}} \times \frac{\text{in}}{2.54\ \text{cm}} = 3.94\ \text{in}$$

b. $\boxed{\text{lbs}} \rightarrow \boxed{\text{g}} \rightarrow \boxed{\text{kg}}$

$$150 \text{ lbs} \times \frac{454 \text{ g}}{\text{lb}} \times \frac{\text{kg}}{1000 \text{ g}} = 68 \text{ kg}$$

c. $\boxed{\text{Å}} \rightarrow \boxed{\text{m}} \rightarrow \boxed{\text{nm}}$

$$10.0 \text{ Å} \times \frac{\text{m}}{10^{10} \text{ Å}} \times \frac{10^9 \text{ nm}}{\text{m}} = 1.00 \text{ nm}$$

d. $\boxed{\text{in}^2} \rightarrow \boxed{\text{cm}^2}$

$$(8.5 \times 11) \text{ in}^2 \times \left(\frac{2.54 \text{ cm}}{\text{in}}\right)^2 = 600 \text{ cm}^2 \text{ or } 6.0 \times 10^2 \text{ cm}^2$$

4. $°C = 77 \text{ K} - 273 = -196 \text{ °C}$
   $°F = 32 + \frac{9}{5}(-196 \text{ °C}) = -321 \text{ °F}$

5. $20 \text{ gal} \times \frac{4 \text{ qt}}{\text{gal}} \times \frac{0.946 \text{ L}}{\text{qt}} \times \frac{1000 \text{ mL}}{\text{L}} \times \frac{0.7 \text{ g}}{\text{mL}} = 53{,}000 \text{ g}$ or 50 kg showing one significant figure.

6. a; 1 mL

7. b; 0.752 g/mL

8. c; dm

9. larger

10. significant figures

Chap

Matter and Energy

A. Outline and Study Hints

Section 2.1

- Matter occupies space and has mass.

Section 2.2

- Energy, the capacity of a system to do work, can be kinetic (motion) or potential. Kinetic energy is proportional to the mass of the moving object and the square of its velocity.

$$KE \propto \text{mass} \times (\text{velocity})^2.$$

Section 2.3

- Mass and the various forms of energy can be transformed, but cannot be created or destroyed.

Section 2.4

- Heat is a the transfer of energy from one object to another. Temperature is the measure of the average kinetic energy of a sample.

Section 2.5

- Calories and joules are both units of heat energy that are used in chemistry. For a given amount of energy added to a substance, the temperature change that results will depend on *how much* substance is being heated.

Section 2.6

- Specific heat is the amount of energy required to raise the temperature of one gram of a substance by 1 degree Celsius. Table 2.3 shows specific heats for several substances. Note that metals have small specific heats. Water has a relatively large specific heat.

$$\text{Specific Heat} = \frac{\text{Kilojoules added or taken away}}{\text{Mass} \times \text{temperature change}}$$

Section 2.7

- Mixtures of matter are either homogeneous (also called solutions) or heterogeneous. Mixtures can be separated through physical processes.
- Pure substances are either compounds or elements.

Section 2.8

- There are only one hundred and nine known elements, each with a one or two letter symbol except the last few elements which have three letter symbols. You should memorize the names and symbols in Table 2.6.
- Elements can be classified as metals and nonmetals and metalloids. The periodic table displays the elements in rows and columns. Nonmetals are on the right side of the table whereas metals are on the left side.

Section 2.9

- The law of definite proportions gives insight to how elements combine to form compounds.

Section 2.10

- Most compounds consist of groups of atoms called molecules. The common diatomic molecules listed in Table 2.9 should be memorized. The chemical formula for a compound shows how many atoms of each element combine to form molecules. The empirical formula shows the simplest ratio of the atoms.
- The chemical formula for the organic solvent benzene is C_6H_6 showing that there are six carbon atoms and six hydrogen atoms connected to form the molecule. The empirical formula is CH showing that the smallest ratio of carbon atoms to hydrogen atoms is one to one.

Section 2.11

- A substance is characterized by its chemical and physical properties.

B. Solutions To In-text Exercises

Exercise 2.1

$$250 \text{ cal} \times \frac{\text{deg} \bullet \text{g}}{\text{cal}} \times \frac{1}{45 \text{ g}} = 5.6 \text{ °C}$$

Exercise 2.2

The temperature change is

$\Delta T = 100 \text{ °C} - 25 \text{ °C} = 75 \text{ °C}$

$$75 \text{ deg} \times \frac{\text{cal}}{\text{g} \bullet \text{deg}} \times 1000 \text{ g } H_2O = 75{,}000 \text{ cal}$$

Exercise 2.3

$$25.0 \text{ g } H_2O \times \frac{\text{cal}}{\text{deg} \bullet \text{g}} \times 16.5 \text{ deg} \times \frac{4.184 \text{ kJ}}{1000 \text{ cal}} = 1.73 \text{ kJ}$$

$$25.0 \text{ g } H_2O \times \frac{\text{cal}}{\text{deg} \bullet \text{g}} \times 41 \text{ deg} \times \frac{4.184 \text{ kJ}}{1000 \text{ cal}} = 4.3 \text{ kJ}$$

Exercise 2.4

$$3.08 \times 10^{-2} \text{ kJ} \times \frac{\text{g} \bullet \text{deg}}{4.18 \times 10^{-3} \text{ kJ}} \times \frac{1}{13.1 \text{ g}} = 0.562 \text{ deg}$$

This is much smaller than the temperature change for the same amount of aluminum because the heat capacity of water is much larger than that of aluminum.

Exercise 2.5

$$\text{Specific heat} = \frac{4.00 \text{ kJ}}{1750 \text{ g} \times (26.10 \text{ °C} - 18.51 \text{ °C})} = 3.01 \times 10^{-4} \text{ kJ/g} \bullet \text{deg}$$

Exercise 2.6

Mg	magnesium	Ta	tantalum	Na	sodium
Mo	molybdenum	K	potassium	Ag	silver
Mn	manganese	Ti	titanium	Fe	iron

Exercise 2.7

barium	Ba	silicon	Si
fluorine	F	arsenic	As

Exercise 2.8

Mg	metal	V	metal
S	nonmetal	C	nonmetal
Ne	nonmetal	P	nonmetal
Cs	metal	I	nonmetal

Exercise 2.9

Na	solid	Br	liquid
Sc	solid	I	solid
He	gas	Hg	liquid
Kr	gas	H	gas

Exercise 2.10

MgO — one magnesium atom, one oxygen atom

Na_2SO_4 — two sodium atoms, one sulfur atom, four oxygen atoms

$Al_2(SO_4)_3$ — two aluminum atoms, three sulfur atoms, twelve oxygen atoms

Exercise 2.11

Chemical formula	Empirical formula
H_2O_2	HO
C_3H_6	CH_2
$C_4H_6O_2$	C_2H_3O

C. Chapter 2 Practice Test:

1. Discuss the energy changes that occur when a bullet is fired into a piece of wood.

2. A glass containing 250 mL of ice water at 0 °C warms to room temperature (25 °C). Calculate the amount of energy absorbed by the water.

3. Without using tables, order the following substances according to increasing specific heats: aluminum, water, ethyl alcohol.

4. Classify the following substances as elements, compounds, homogeneous mixtures (solutions), or heterogeneous mixtures: wood, magnesium, propane, baking soda, steel.

5. Write chemical symbols for the following elements: iron, carbon, copper, phosphorus, sodium.

6. Name the elements and the number of atoms of each element for the following formulas:
   Epsom salts, $MgSO_4$, is a hydrate; see text page 371.
   hydrochloric acid, HCl
   bleach, NaClO
   carbon tetrachloride, CCl_4
   magnesium hydroxide, $Mg(OH)_2$

7. Si is classified as a
   a. metal b. nonmetal c. metalloid d. none of these

8. The fact that water is always composed of 11% hydrogen and 89% oxygen is an example of the law of ________.

9. Two elements may combine in (only one/ more than one) set of proportions.

10. All samples of a given pure substance
    a. contain no chemicals
    b. are heterogeneous
    c. have identical compositions
    d. have variable compositions

D. Answers To Practice Test

1. The bullet initially has chemical potential energy which is converted into kinetic energy as it is fired. The chemical reaction that produces the kinetic energy is the explosion of the powder in the shell. The kinetic energy is transformed into heat after the lead bullet has struck the object.

2. $25 \text{ deg} \times \frac{\text{cal}}{\text{g} \bullet \text{deg}} \times 250 \text{ mL} \times \frac{\text{g}}{\text{mL}} = 6300 \text{ cal}$

3. Water has highest specific heat, aluminum the lowest, and ethyl alcohol is intermediate.

4. Wood is a heterogeneous mixture of mostly organic compounds. Magnesium is a metallic element. Propane is flammable compound that usually has an odorous additive for safety reasons. Baking soda is a compound, sodium hydrogen carbonate. Steel is a homogeneous mixture (solution) of metals, mostly iron.

5. Fe, C, Cu, P, Na

6. $MgSO_4$ — one magnesium atom; one sulfur atom; four oxygen atoms

   HCl — one atom each of hydrogen and chlorine

   $NaClO$ — one atom each of sodium, chlorine and oxygen

   CCl_4 — one atom of carbon; four chlorine atoms

   $Mg(OH)_2$ — one magnesium atom; two each of oxygen and hydrogen.

7. c

8. definite proportions

9. more than one

10. c

Chapter 3

The Atomic Theory

A. Outline and Study Hints

Section 3.1

- Although Democritus perhaps conceived of the idea of atoms, the development of atomic theory required the quantitative work of the nineteenth century.

Section 3.2

- Dalton's theory, outlined in 7 statements, should be memorized. Essentially, his theory states that matter consists of atoms. The atoms of different elements differ in mass and size, whereas atoms of a given element are identical.

Section 3.3

- There are two forms of electric charge; positive (+) and negative (-).

Section 3.4

- Electrons are negatively charged particles found in all atoms.

Section 3.5

- Protons are positively charged particles, 1840 times heavier than electrons, and are also found in all atoms.

Section 3.6

- Rutherford's experiment suggested that atoms are made of electrons surrounding a positively charged nucleus.

Section 3.7

- The number of protons in an atom, called the atomic number, uniquely characterizes an element.

Section 3.8

- Neutrons are neutrally charged particles that are found in the nucleus. The sum of the atomic number and the number of neutrons is called the mass number. Isotopes of a given element differ in mass number.

B. Solutions to In-Text Exercises

Exercise 3.1

^{28}Si has an atomic number of 14 and a mass number of 28, thus the number of neutrons is 28 - 14 = 14.

^{29}Si 29 - 14 = 15 neutrons.

^{30}Si 30 - 14 = 16 neutrons.

Exercise 3.2

^{35}Cl has an atomic number of 17 and a mass number of 35, thus the number of neutrons is 35 - 17 = 18.

^{14}C has an atomic number of 6 and a mass number of 14, thus the number of neutrons is 14 - 6 = 8.

C. Chapter 3 Practice Test

1. The charge of an electron is
   a. positive.
   b. negative.
   c. zero.
   d. dependent on the atom.

2. Atoms
   a. contain electrons.
   b. are mostly space.
   c. are neutrally charged.
   d. are all of the above.

3. Protons and neutrons
   a. have about the same mass.
   b. have the same charge.
   c. have the same color.
   d. occur in equal numbers in the nucleus of atoms.

4. The atomic number represents the number of
   a. electrons.
   b. protons.
   c neutrons.
   d. bevatrons.

5. Ions have
   a. equal numbers of electrons and protons.
   b. more electrons than protons.
   c. more protons than electrons.
   d. unequal numbers of electrons and protons.

6. Which element commonly has no neutrons in its nucleus?

7. The element with atomic number 51 is _______.

8. Who showed that an atom's mass is concentrated in its nucleus?

9. Magnesium has ________ protons in its nucleus.

10. Isotopes of the same element have the same number of _______ but differ in the number of ______.

D. Answers To Practice Test:

1. b

2. d

3. a

4. b

5. d

6. hydrogen, H

7. antimony, Sb

8. Rutherford

9. 12

10. protons, neutrons

Chapter 4

Electron Structure of the Atom

A. Outline and Study Hints

Section 4.1

- Electrons in atoms are characterized by their energy levels which can be measured by spectroscopy. Electron energy is quantized.

Section 4.2

- The classical Bohr theory of atoms has been replaced by a wave mechanics theory of electron behavior.

Section 4.3

- There are four quantum numbers that constitute a complete description of the energy of an electron within an atom. These can be thought of as the address of the electron. The complex rules and patterns for these numbers should be committed to memory.
- Do not try to rationalize these rules.
- The principle quantum number, also called the main level, is represented by **n**. The number of and designations of the sublevels within each main level are given in Table 4.1. The number of orbitals within each sublevel are shown in Table 4.2.
- Two electrons can be contained within an orbital.

Section 4.4

- Figure 4.7 shows the order of energies of the sublevels, for example sublevel 4*s* is lower in energy that 3*d*. You should learn this order.

Section 4.5

- The aufbau principle and Hund's rule are used to assign electrons to sublevels and orbitals within atoms. The energy sublevels are filled first, with no more than two electrons in each orbital.
- For a given sublevel, electrons will occupy separate orbitals if possible. The electron configuration is a shorthand way of indicating which electrons occupy which sublevels.

Section 4.6

- An electron probability plot is a representation showing where the electron in a given sublevel or orbital is more likely to be found.

Section 4.7

- Most properties of atoms are due to the outer electrons also called valence electrons. A Lewis diagram shows these electrons as dots around the chemical symbol.

Section 4.8

- No atom can hold more than eight valence electrons. This octet of electrons will become an important concept when we study the interaction between atoms.

B. Solutions to In-Text Exercises

Exercise 4.1

The fourth main energy level,

n = 4 holds $2\,n^2 = 2 \times 4^2 = 2 \times 16 = 32$ electrons.

The *s* sublevel holds 2 electrons, the *p* sublevel can have as many as 6 electrons, *d* as many as 10 and *f* as many as 14 electrons.

Exercise 4.2

$4d^3$ pronounced " four dee three ".

Exercise 4.3

$3d^6$ is the notation that indicates six electrons in the *d* sublevel of the n = 3 main level.

Exercise 4.4

S has atomic number 16, thus S will have this many electrons in a neutral atom. The boxes represent orbitals and the arrows represent electrons within those orbitals.

[↑↓] [↑] [↑]	3*p*
[↑↓]	3*s*
[↑↓] [↑↓] [↑↓]	2*p*
[↑↓]	2*s*
[↑↓]	1*s*

Exercise 4.5

Mg, magnesium has 12 electrons.

Exercise 4.6

Zr has 40 electrons, thus using the aufbau principle the electron configuration would be

$1s^2\,2s^2\,2p^6\,3s^2\,3p^6\,4s^2\,3d^{10}\,4p^6\,5s^2\,4d^2$.

Exercise 4.7

The electron configurations of V (23 electrons) and Br (35 electrons) are

V $1s^2\,2s^2\,2p^6 3s^2\,3p^6\,4s^2\,3d^3$

Br $1s^2\,2s^2\,2p^6\,3s^2\,3p^6\,4s^2\,3d^{10}\,4p^5$

Exercise 4.8

The electron configuration is used to deduce the number of valence electrons. Because phosphorus has 15 electrons, then

$1s^2\ 2s^2\ 2p^6\ 3s^2\ 3p^3$

is the electron configuration. Note that there are 5 valence electrons having the highest principle quantum number ($3s^2$ and $3p^3$).

Thus the Lewis diagram is

$$\overset{\bullet\bullet}{\bullet\,\text{P}\,\bullet}\atop\bullet$$

Exercise 4.9

N nitrogen, P phosphorus, and As arsenic all have 5 valence electrons.

C. Chapter 4 Practice Test

1. De Broglie wavelength is
   a. the distance between electrons.
   b. applicable to moving particles such as electrons.
   c. important in the Bohr atomic model.
   d. one of the four quantum numbers.

2. Explain the following notation: $3p$.

3. Which of the following sublevels is not allowed according to the quantum rules?
   a. $1s$
   b. $2d$
   c. $3s$
   d. $3d$

4. Write the orbital distribution diagram for potassium.

5. Write the electron configuration for aluminum.

6. Draw the Lewis diagram for aluminum.

7. The electron configuration for Be is
   a. $1s^2\ 2s^2$
   b. $1s^2\ 2s^2\ 2p^5$
   c. $1s^2\ 2p^2$
   d. $1s^2\ 2p$

8. What element has the electron configuration $1s^2\ 2s^2\ 2p^3$?

9. How many electrons are there in the outer energy level of a sulfur atom?

10. Lewis diagrams show ________ electrons as dots.

D. Answers To Practice Test:

1. b

2. This represents the "three pee" energy sublevel within the n = 3 main level. This sublevel has three orbitals.

3. b, the n = 2 main level has only an *s* and a *p* sublevel.

4. K has 19 electrons.

[↑]	4*s*
[↑↓] [↑↓] [↑↓]	3*p*
[↑↓]	3*s*
[↑↓] [↑↓] [↑↓]	2*p*
[↑↓]	2*s*
[↑↓]	1*s*

5. Al has 13 electrons.
   $1s^2\ 2s^2\ 2p^6\ 3s^2\ 3p^1$

6. Al has 3 valence electrons ($3s^2$ and $3p^1$) thus the Lewis formula is
   $\cdot\dot{\mathrm{Al}}\cdot$

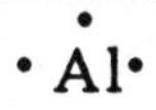

7. a

8. nitrogen

9. six

10. valence or outer

Chapter 5

Families of Elements and the Periodic Table

A. Outline and Study Hints

Section 5.1

- The historical development of the modern periodic table involved Newlands from England and Mendeleev from Russia. Mendeleev arranged the elements in order of increasing atomic mass, but then he rearranged the order according to known chemical and physical properties. His prediction of the missing elements gallium, scandium and germanium and their subsequent discovery a few years later solidified his now famous connection to the periodic table. You should memorize the typical shape of the table:

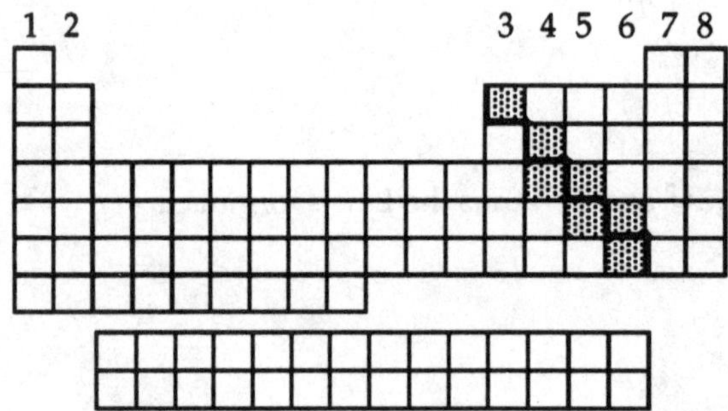

- Note that Roman numerals such as IA, IIA are alternatively used instead of Arabic numerals 1, 2.

Section 5.2

- The seven horizontal rows in the periodic table are called periods. The vertical columns are called families or groups.
- The first and last families are respectively the alkali metals and the noble gases. You should be able to identify the representative elements, the transition elements, the lanthanides, and the actinides (Figure 5.1) on the periodic table.

Section 5.3

- Although originally based on trends in physical and chemical properties, the periodicity of the elements is also reflective of the electron configurations. Comparison of electron configurations reveals the similarity of elements within a family. For example, all of the alkali metals (group I) have a single *s* electron in the last sublevel, whereas the alkaline earth metals (group II) all have two valence electrons in an *s* sublevel. These two families are called the *s* block elements. The *p* block elements, which contain the halogens (column 7) and the noble gases (column 8) and the *d* block elements (transition metals) have similar electron configuration correlations within each family. In each case, we see that atoms within a family have the same number of valence electrons in the same type of sublevels.
- This correlation allows us to write electron configurations by noting the location of an atom within the periodic table. As an example, notice that silicon is in the second column of the third period of the *p* block. Thus there are two electrons in the $3p$ sublevel. The other sublevels before this are completely filled, thus the electron configuration of Si is $1s^2\ 2s^2\ 2p^6\ 3s^2\ 3p^2$.

Section 5.4

- The most useful numbering system for families uses Roman numerals (I through VII) followed by letter A for the representative elements and B for the transition metals.
- The number of valence electrons is the group number for the A elements. Silicon is in group IVA and thus has four valence electrons.

Section 5.5

- The most metallic elements are found in the lower left of the periodic table.
- The metalloids divide the metals from the nonmetals in the upper right of the periodic table.

Section 5.6

- Atomic size trends are correlated with the periods and groups in the periodic table. Size decreases as you proceed across a period from left to right. This is due to the increasing nuclear charge pulling the electrons closer to the nucleus.
- As you proceed down a family, the addition of a complete level of electrons causes the size of atoms to increase.

Section 5.7

- Energy is needed to remove an electron from an atom or an ion. The removal process is called ionization.
- The ionization energy tends to increase as the size of the atom or ion decreases. Thus ionization energy decreases as you go down a group and increases from left to right across a period of the periodic table.
- As a whole metals tend to have small ionization energies and nonmetals tend to have large ionization energies.

Section 5.8

- Electron affinity is a measure of the attraction of an atom for an electron. The trend for electron affinities follows that of ionization energies; atoms that are difficult to ionize readily accept an electron.

Section 5.9

- The similarity of electron configurations for elements within a family cause the elements to have similar chemical and physical properties.
- The alkali metals have low melting and boiling points. They are the most reactive metals, followed by the alkaline earth metals.
- The halogens are the most reactive nonmetals. Most are gases at room conditions. Bromine is a brown liquid and iodine is a purple shiny solid.
- The noble gases are essentially nonreactive.
- The transition elements are all metals whose physical and chemical properties are similar.

Section 5.10

- The variable distance of valence electrons from the nucleus causes subtle differences in the properties of elements within a family. Metallic character and size increases as you go down a family. Metals tend to be more reactive as you go down the family whereas nonmetals tend to be less reactive.

Section 5.11

- Hydrogen is unique among elements and is not easily placed within a family. This is due to its small size and the fact that its single electron is not shielded by inner electrons.

B. Solutions to In-Text Exercises

Exercise 5.1

The first two letters in in an element's name are usually a useful guide to its symbol. The atomic symbol for silicon is Si, atomic number 14, found in period 3 (third row).

Exercise 5.2

Magnesium, atomic number 12, is in period 3, in the *s* block. The last two electrons are thus in the 3*s* sublevel; lower energy sublevels are filled. Thus the electron configuration is
$1s^2\ 2s^2\ 2p^6\ 3s^2$

Exercise 5.3

The number of valence electrons is given by the family number for the representative elements, thus

Element	Family Number	Valence Electrons
Al	IIIA	3
Se	VIA	6
Xe	VIIIA	8
F	VIIA	7
He	VIIIA	2
Ge	IVA	4

He, helium, has only two electrons therefore both of these are valence electrons. Helium is placed with the noble gases because it is unreactive. Some periodic tables place He in family IIA because of its similar electronic configuration to that of alkaline earth metals.

Exercise 5.4

Rubidium is an alkali metal like sodium and cesium and iodine is a halogen like chlorine and bromine. Thus we expect that Rb and I would combine to form a compound is similar to other alkali metals and halogens, RbI.

Exercise 5.5

Helium is the first member of the noble gas family. We expect it to be small, nonreactive and possess no metallic characteristics.
Aluminum is the second member of family IIIA. We expect aluminum atoms be larger than boron atoms and more metallic. It is also rather reactive as a metal.

C. Chapter 5 Practice Test

1. Mendeleev is most well known because
   a. he invented the idea of a periodic table.
   b. he was an avid balloonist.
   c. he invented three elements.
   d. he predicted the existence of three elements.

2. The periodic table is arranged in rows called ______ and columns called _______ or _______ .

3. Which elements are located in the fourth period and groups IIIA through VA?

4. The periodic table has
   a. regions known as blocks.
   b. blocks and neighborhoods.
   c. transition metals in the *p* block.
   d. an alphabetical order.

5. Each of the representative elements has a number of valence electrons equal to the ______.

6. Write the electron configuration for arsenic by referring only to a periodic table.

7. Which element has the largest atoms?
   a. oxygen
   b. sulfur
   c. bromine
   d. iodine

8. Which element is a metalloid?
   a. metalonium
   b. radium
   c. tellurium
   d. carbon

9. Explain why the ionization energy of a metal like sodium is less than that of a nonmetal like chlorine.

10. Predict the formula for a compound containing sulfur and aluminum given that the following aluminum compounds are known to exist: $AlCl_3$, Al_2O_3, AlP.

D. Answers To Practice Test:

1. d

2. periods, families, groups

3. Ga, Ge, As

4. a

5. group or family number

6. $1s^2\ 2s^2\ 2p^6\ 3s^2\ 3p^6\ 4s^2\ 3d^{10}\ 4p^3$

7. d

8. c

9. Sodium has only one valence electron whereas chlorine has seven. Sodium atoms are larger, thus it takes less energy to remove the electron from a sodium atom than a chlorine atom.

10. Sulfur is in the same group as oxygen thus the formula for aluminum sulfide would be similar to Al_2O_3; Al_2S_3.

Chapter 6

How Elements Form Compounds

A. Outline and Study Hints

Section 6.1

- An atom has equal numbers of positively charged protons and negatively charged electrons.
- When electrons are lost or removed from an atom, a positively charged cation results.
- When electrons are gained or received by an atom, a negatively charged anion results.
- Ions are designated with chemical symbols and the charge is shown as a right hand superscript. An example is the hydrogen ion, written H^{1+} or H^{+}.
- Metal atoms often lose electrons to form cations whose electron configurations are like those of noble gases; nonmetal atoms often gain electrons to form anions whose electron configurations are like those of noble gases. We say the electron configurations are isoelectronic with those of noble gases.
- This generality is called the octet rule. Noble gas atoms are unreactive because they already have 8 valence electrons.
- Cations are smaller than the atoms from which they are formed while anions are larger than the atoms from which they are formed.

Section 6.2

- When electrons are transferred from a metal to a nonmetal we say that the metal has been oxidized and the nonmetal has been reduced. These reactions are called electron transfer or oxidation-reduction.
- Ions that are produced are attracted to form ionic compounds held together with ionic bonds. Most ionic compounds separate or dissociate into cations and anions when dissolved in aqueous (water) solutions.

Section 6.3

- The formulas for ionic compounds are deduced by requiring that sufficient cations and anions combine to form a neutral formula.
- An example is aluminum sulfide which consists of aluminum ions and sulfide ions

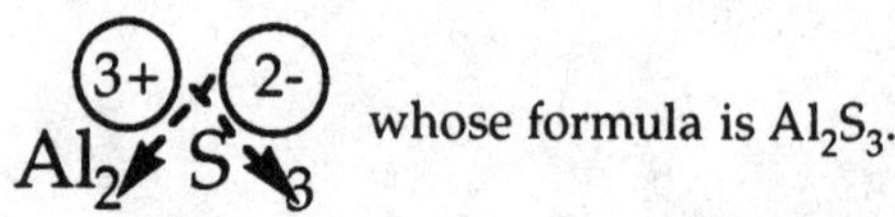

whose formula is Al_2S_3.

- Notice that the formula subscripts are obtained by using the charge of the other ion as shown in the diagram. The subscript 3 for the sulfur atom is the same as the charge on the aluminum ion and the subscript 2 on the aluminum atom is the same as the charge on the sulfide ion.

Section 6.4

- Atoms may share valence electrons between them to satisfy the octet rule. This idea is conveniently displayed in Lewis diagrams which show shared electron pairs, called covalent bonds, between atoms as dots or lines. Diatomic chlorine is represented as

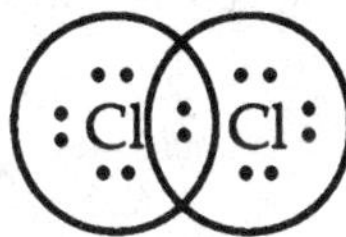

- Each circle around the atom encloses eight valence electrons, showing that the octet rule is satisfied.

Section 6.5

- Covalent bonds are made up of shared electrons between atoms, but only when the atoms are identical are the electrons shared equally. In most cases, the electrons are pulled closer to the more electronegative atom and a polar covalent bond results.
- The most electronegative elements are the nonmetals in the upper right hand corner of the periodic table; the least electronegative elements are found in the bottom left hand side of the periodic table.
- If the difference in electronegativity is larger than about 1.9, then an ionic bond results because electrons are transferred rather than shared.

Section 6.6

- There are several notations employed in drawing covalent molecules. Thus for Cl_2 we could write

:Cl:Cl: — Lewis dot diagram

×Cl×Cl: — Showing electron contribution from each atom

Cl—Cl — Bond line diagram

- We will often use bond line diagrams for our molecules. Each dash line between the two elements represents two electrons.

Section 6.7

- Many atoms form several bonds and a good rule is that the number of bonds is equal to eight minus the group number of the element. Thus oxygen is expected to form 8 - 6 = 2 bonds and nitrogen would commonly form 8 - 5 = 3 bonds.

Section 6.8

- Many atoms form multiple bonds, called double or triple bonds such as

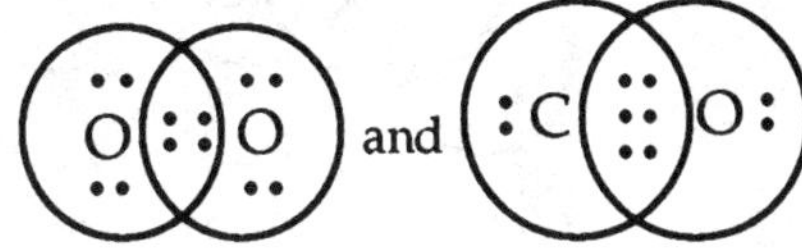

where the circles enclose 8 valence electrons.

- Coordinate covalent bonds form when the shared electrons originate from only one of the atoms.

Section 6.9

- **The systematic method used to draw Lewis diagrams follows these steps which we will apply to formaldehyde, CH_2O.**

(1) Draw a preliminary structure with the central atom surrounded by other atoms.

H
C O
H

(2) Determine the number of bonds within the molecule using the formula:

$$\begin{pmatrix}\text{number of atoms}\\ \text{times eight}\\ \text{(or two for H)}\end{pmatrix} - \begin{pmatrix}\text{sum of valence}\\ \text{electrons from}\\ \text{each atom}\end{pmatrix} = \begin{pmatrix}\text{number of}\\ \text{electrons that}\\ \text{must be shared}\end{pmatrix}$$

$$(2 \times 2 + 2 \times 8) - (1 + 1 + 4 + 6) = 8 \text{ electrons (4 bonds)}$$

(3) Draw in the bonds.
Because oxygen forms two bonds, we use a double bond between carbon and oxygen.

H
:C::O
H

(4) Draw in remaining electrons to satisfy the octet rule.
We have used 8 of the 12 valence electrons available, thus we place the four remaining electrons on oxygen to complete its octet

H
:C::Ö:
H

(Circles are for clarification only and do not need to be included in your answers.)

(5) Check and count the electrons.
We see that we have used all 12 valence electrons and the octet rule is satisfied by all atoms as the circles indicate.

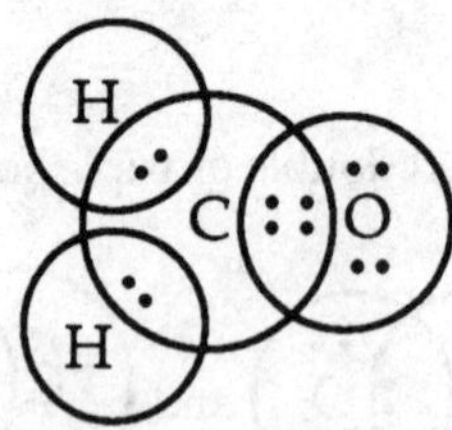

We could equivalently write

```
H        ..
  \     
   C = O        or        :O:
  /     ..                 ||
H                      H — C — H
```

as the location of atoms and electrons is not important at this point.

(6) Some atoms may have more than eight electrons. These will be in periods 3 or higher. Such large atoms can accommodate more than 8 valence electrons.

Section 6.10

- Lewis diagrams may be drawn for polyatomic ions. Anionic charges are added to the number of valence electrons whereas cationic charge is subtracted from the number of valence electrons as shown in these examples:

Ion	Number of valence electrons
SO_4^{2-}	$6+4\times6+2=32$
$CH_3NH_3^+$	$4+3\times1+5+3\times1-1=14$

Remember that an atom has its group number of valence electrons.

Section 6.11

- Formulas for compounds containing polyatomic ions treat the polyatomic ions as separate entities, often enclosed in parentheses. An example is aluminum sulfate

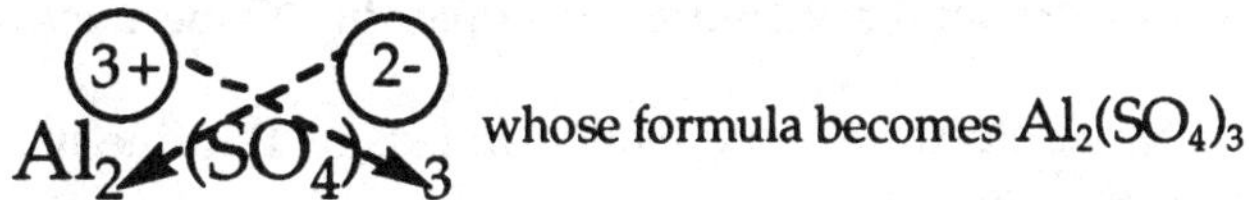

where parentheses have enclosed the sulfate ion to avoid confusion.

Section 6.12

- Molecules have three dimensional shapes.
- Note that ammonia, NH_3 and methane, CH_4, are not planar but instead are shaped like pyramids, see page 180 of text.

Section 6.13

- Knowing the shape as well as the bond polarities allows us to predict whether molecules are polar. We say such molecules possess a dipole moment.
- Highly symmetric molecules such as CH_4 are usually nonpolar, because C-H dipoles cancel each other.

B. Solutions to In-Text Exercises

Exercise 6.1

O^{2-}; the superscript, 2-, shows that oxygen has gained two electrons.

Exercise 6.2

Cl is likely to gain one electron so that it will then have as many electrons as argon.

Exercise 6.3

Al will ordinarily lose three electrons so that it would then have

$$Al \rightarrow Al^{3+} + 3e^-$$

(3 valence electrons) $\cdot\dot{Al}\cdot \rightarrow Al^{3+}$ (no valence electron)

Exercise 6.4

Li loses 1 electron, O gains 2 electrons; Li_2O,

Sr loses 2 electrons, Cl gains 1 electron; $SrCl_2$

Al loses 3 electrons, F gains 1 electron; AlF_3

Exercise 6.5

Oxygen will more readily attract the shared electron pair in CO, because oxygen has a higher electronegativity.
Fluorine attracts the shared electrons more readily in BrF, as fluorine has the highest electronegativity of all the elements.

Exercise 6.6

Reacting elements	Electronegativities of these elements	Difference in electronegativities	Kind of bond formed
Cs + F	Cs, 0.7; F, 4.0	3.3	Ionic
O + Na	O, 3.5; Na, 0.9	2.6	Ionic
C + H	C, 2.5; H, 2.1	0.4	Covalent
Si + Cl	Si, 1.8; Cl, 3.0	1.2	Covalent

Note that if the difference in electronegativities is greater than 1.9, then the bond is ionic.

Exercise 6.7

H:F: , :I:I: , :I:Br:

H—F| , |I—I| , |I—Br|

Exercise 6.8

```
  ••          Cl
 :Cl:          |
  ••  ••       P— Cl
 :P:Cl:        |
  ••  ••       Cl
 :Cl:
  ••
```

Exercise 6.9

```
                      H
    H                 |
    ••           H — C — H
 H:C:H                |
    ••                H
    H
```

Exercise 6.10

Electron dot drawing of H_2O_2

```
     ••    ••
 H — O — O — H
     ••    ••
```

Exercise 6.11

Electron dot drawing of CO_2

```
 ••       ••           ••
 O = C = O   or   :O — C ≡ O:
 ••       ••           ••
```

Exercise 6.12

Electron dot drawing of H_3PO_4

```
             ••
            :O:
             |
     ••             ••
 H — O  —  P  —  O — H
     ••             ••
             |
            :O:
             |
             H
```

Exercise 6.13

Electron dot drawing of ClO_4^- and NO_3^-

```
 [        ••        ]-      [        ••    ]-
 [       :O:        ]       [       :O:    ]
 [        |         ]       [        |     ]
 [  ••          ••  ]       [  ••          ]
 [ :O  — Cl —  O:   ]       [  O = N       ]
 [  ••          ••  ]       [  ••          ]
 [        |         ]       [        |     ]
 [       :O:        ]       [       :O:    ]
 [        ••        ]       [        ••    ]
```

Exercise 6.14

$Al(BrO_3)_3$, Rb_2SO_4

Exercise 6.15

$Ca_3(PO_4)_2$, $SrSO_4$

Exercise 6.16

The Lewis diagram for HNO_3 shows that hydrogen has a partial positive charge and the nitrate ion has a partial negative charge.

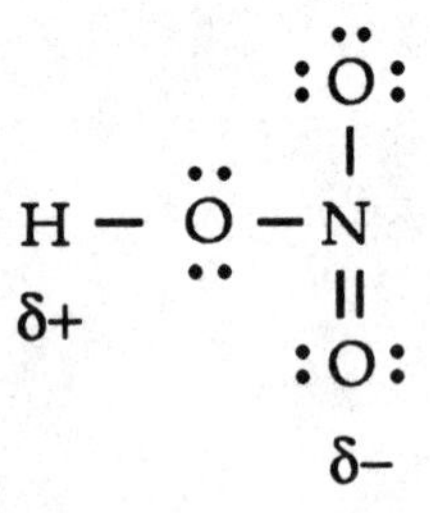

```
            ..
           :O:
      ..    |
  H — O — N
      ..    ||
  δ+       :O:

            δ–
```

Thus HNO_3 is polar.

C. Chapter 6 Practice Test

1. Which of the following ions has a noble gas electron configuration?
   a. Ca^+
   b. Sb^{2+}
   c. Te^{2-}
   d. Al^{4+}

2. Write the electron configuration of Mg^{2+}.

3. When a metal __________ electrons the ion formed is called a __________ .

4. Predict the formulas of the ions of bromine, strontium and sulfur.

5. Which of the following is the correct formula for the ionic compound formed when barium reacts with iodine?
   a. BI_3
   b. BaI
   c. Ba_2I
   d. BaI_2

6. What ions form when Cs_3P dissolves in water?

7. Which of the following would be ionic compounds?
   a. CCl_4
   b. CO
   c. LiF
   d. CaS

8. Write Lewis formulas for C_2H_6, SO_2, HCO_2H (challenging).

9. Write the formula of the ionic compound that is predicted for NH_4^+ and PO_4^{3-}.

10. Molecules that have ________ bonds can have dipoles; some ________ molecules have polar bonds but their ________ causes the dipoles to cancel.

D. Answers To Practice Test:

1. c

2. same as neon; $1s^2\,2s^2\,2p^6$.

3. loses, cation

4. Br^-, Sr^{2+}, S^{2-}

5. d

6. Cs^+ and P^{3-}

7. c and d

8.

```
    H   H
    |   |                              :O:
H - C - C - H                           ||
    |   |        ..    ..    ..               ..
    H   H   ,    O  =  S  -  O:  ,  H - C  -  O - H
                 ..          ..               ..
```

9. $(NH_4)_3PO_4$

10. polar, nonpolar, shape or geometry.

Chapter 7

An Introduction to Naming Compounds

A. Outline and Study Hints

Section 7.1

- Remember that oxidation-reduction reactions involve the transfer of electrons. We use oxidation numbers also called oxidation states, to keep track of transferred electrons.
- You should memorize the arbitrary rules for assigning oxidation numbers to atoms in compounds:

1. Atoms in pure elements have oxidation numbers of zero.
2. The alkali metals Na through Cs are assigned +1 oxidation number in compounds.
3. The alkaline earth metals Be through Ba are assigned +2 oxidation number in compounds.
4. The common oxidation numbers of metals Al^{3+}, Ag^{+}, Cd^{2+}, Zn^{2+} are shown by their charges.
5. The oxidation number of hydrogen is usually +1 (except in hydrides where it is -1) .
6. The oxidation number of oxygen is -2, except in OF_2 and peroxides.
7. Halogens have oxidation states of -1.
8. The sum of oxidation states of atoms equals the charge on the species (ion or molecule).
9. Electrons are assigned to the more electronegative atom in a covalent bond.

- In applying the above rules, you should go in the order shown. Note that atoms, not compounds, are assigned oxidation numbers. Nonmetals and transition metals have variable oxidation numbers dependent on the particular compound.

Section 7.2

- Binary compounds contain only two elements. Binary metallic formulas always begin with the metal. Names of binary compounds state the elements in the order of the formula and always end in **-ide**.
- Naming binary metallic compounds that contain an alkali metal, an alkaline earth metal, aluminum, silver, cadmium or zinc, you simply state the metal name, followed by the nonmetal name ending in **-ide**.
- For binary metallic compounds containing a metal with a variable oxidation state, name the metal followed by the oxidation number as a Roman numeral in parentheses after the metal. This is called the Stock or IUPAC system. An older system is sometimes used. The higher oxidation state has an **-ic** ending whereas the lower oxidation number uses an **-ous** ending. Table 7.2 should be memorized.
- Mercury is peculiar; its two ions are listed below with Stock and old names shown.

mercury(I)	mercurous	Hg_2^{2+}
mercury(II)	mercuric	Hg^{2+}

- Binary nonmetallic compounds use Greek prefixes listed in Table 7.3 to indicate the number of each element in the formula unit. The prefix **mono-** is not used for the first element in the compound name. Note that these prefixes are not used in naming binary metallic compounds. You should probably make an effort to memorize most of the representative element chemical symbols if you haven't already.

Section 7.3

- Binary nonmetallic compounds that contain hydrogen form acidic solutions when dissolved in water. We distinguish these aqueous acids from the binary compounds by substituting the prefix **hydro-** for hydrogen and ending with **-ic acid** rather than **-ide**. Thus hydrogen chloride is HCl(g), whereas HCl(aq) is hydrochloric acid. You must know oxidation numbers to deduce formulas from names. Hydrosulfuric acid consists of hydrogen ions (H^+) and sulfide ions (S^{2-}), thus the formula is H_2S(aq).

Section 7.4

- Polyatomic names and charges in Table 7.4 must be memorized. The order for the atoms is sometimes reverse, for example hypochlorite can be written as OCl^- or ClO^-. Acetate is sometimes written as CH_3COO^- as well as $C_2H_3O_2^-$.
- Some hydrogen containing ions are

bicarbonate	HCO_3^-	hydrogen carbonate
bisulfate	HSO_4^-	hydrogen carbonate
bisulfite	HSO_3^-	hydrogen sulfite

- There is a pattern in the nonmetal oxygen-containing polyatomic names. The -ate ions (arsenate, chlorate, iodate, nitrate, and phosphate) are used to derive other ions by addition or subtraction of oxygen. Chlorate serves as an example:

$$\overset{+7}{ClO_4^-} \xleftarrow{+O} \overset{+5}{ClO_3^-} \xrightarrow{-O} \overset{+3}{ClO_2^-} \xrightarrow{-O} \overset{+1}{ClO^-}$$

perchlorate | chlorate | chlorite | hypochlorite

- Note that the oxidation number of chlorine varies in this series; perchlorate has highest and hypochlorite the lowest oxidation number of chlorine.
- An efficient way to learn the names of these (and other) ions is to write the names and formulas on opposite sides of 3 × 5 cards. You can carry these with you and quiz yourself while waiting in the line at the bank for instance.
- To name compounds containing polyatomic ions place the cation first then the anion. You may want to write the charges next to the ions to help you write the formulas. For example $Cr_2(SO_3)_3$ consists of ions:

Cr^{3+}	SO_3^{2-}
chromium(III) or chromic	sulfite

- Table 7.2 in the text has an error in the *Old Names* of chromium. Chromous is the old name for chromium (II) and chromic is the old name for chromium (III).
- The charge on the chromium is deduced because $2x + 3(-2) = 0$, the charge on the formula unit. The name of this compound is either chromium(III) sulfite or chromic sulfite.

- **Writing the formula from the name is usually easier. For example iron(II) perchlorate consists of, Fe^{2+} and ClO_4^- thus the formula has two ClO_4^- ions for every Fe^{2+} ion: $Fe(ClO_4)_2$. The parentheses are useful for clarity.**
- **Hydrogen cations with polyatomic anions are acids when dissolved in aqueous solutions. When the anion contains oxygen, the name of the acid follows the rules illustrated by the example involving chlorate.**

anion	name	acid	name
ClO_4^-	<u>per</u>chlor<u>ate</u>	$HClO_4$	<u>per</u>chlor<u>ic</u> acid
ClO_3^-	chlor<u>ate</u>	$HClO_3$	chlor<u>ic</u> acid
ClO_2^-	chlor<u>ite</u>	$HClO_2$	chlor<u>ous</u> acid
ClO^-	<u>hypo</u>chlor<u>ite</u>	$HClO$	<u>hypo</u>chlor<u>ous</u> acid

Section 7.5

- **If a compound has more than one type of cation, Greek prefixes are used to designate the number of each type. The number of anions is deduced by charge neutrality. The compound KRb_2PO_4 contains K^+, Rb^+, and PO_4^{3-} ions. It is called potassium dirubidium phosphate.**

Section 7.6

- **Some common names of compounds are worth knowing such as:**

H_2O	water	$NaHCO_3$	baking soda
NH_3	ammonia	C_2H_2	acetylene
NaOH	lye	NaCl	table salt
CaO	lime		
$CaSO_4 \bullet 2\ H_2O$	gypsum	($\bullet 2\ H_2O$ means $(H_2O)_2$)	

B. Solutions to In-Text Exercises

Exercise 7.1

In PBr_3, Br, a halogen, thus is -1 and P is thus +3 because the sum [+3 + 3(-1)] must be equal to zero. In PBr_5, Br is -1 and P is +5.

Exercise 7.2

In $CHCl_3$, H is +1 (rule 5), Cl is -1 (rule 7) and C is +2 (rule 8). Note that rule 9 implies that H is -1 because C is more electronegative than H; however, rule 5 takes precedence over rule 9.

Exercise 7.3

In NO_3^-, O is -2 (rule 6), and N is +5 (rule 8; $x + 3(-2) = -1$ thus $x = +5$).

Exercise 7.4

In NO_2^-, O is -2 (rule 6), and N is +3 (rule 8).

Exercise 7.5

BaO is barium oxide. K_3N is potassium nitride. Al_2S_3 is aluminum sulfide . Note that the last few letters of the nonmetal have been deleted and replaced by -ide..

Exercise 7.6

Silver iodide is AgI and zinc oxide is ZnO. Recall the charges on silver and zinc are (Ag^+, Zn^{2+}).

Exercise 7.7

CoSe is cobalt(II) selenide. $CrBr_2$ is chromium(II) bromide. Hg_2Cl_2 is mercury(I) chloride. Note that the nonmetal oxidation state is used to deduce the oxidation state on the metal. In Hg_2Cl_2, the chlorine has an oxidation number of -1, thus Hg_2^{2+} must be the metal ion, called mercury(I).

Exercise 7.8

Iron(II) oxide is FeO (oxide is O^{2-} and iron(II) is Fe^{2+} which are combined to formFeO), and tin(IV) fluoride is SnF_4.

Exercise 7.9

PbO_2 is plumbic oxide (Pb^{4+} and two O^{2-}). CuI_2 is cupric iodide (Cu^{2+} and two I^-). Hg_2Cl_2 is mercurous chloride (Hg_2^{2+} and two Cl^-).

Exercise 7.10

Ferrous chloride is (Fe^{2+} and two Cl^- ions) is $FeCl_2$ and stannic fluoride (Sn^{4+} and four F^- ions) is SnF_4.

Exercise 7.11

SO_2 is sulfur dioxide. SO_3 is sulfur trioxide. S_2O_3 is disulfur trioxide.

Exercise 7.12

Selenium trioxide is SeO_3 and silicon tetrafluoride is SiF_4.

Exercise 7.13

HI is hydroiodic acid and H_2Se is hydroselenic acid (binary acids begin with hydro- and end with -ic acid).

Exercise 7.14

Hydrofluoric acid is HF (hydrogen is H^+, fluoride is F^-).

Exercise 7.15

$NaClO_2$ is sodium chlorite and CaC_2O_4 is calcium oxalate.

Exercise 7.16

Cu_2SO_4 is copper(I) sulfate and $Cr_2(SO_3)_3$ is chromium(III) sulfite.

Exercise 7.17

Cobalt(III) acetate is $Co(C_2H_3O_2)_3$.

Exercise 7.18

Cu_2SO_4 is cuprous sulfate and $SnSO_4$ is stannous sulfate.

Exercise 7.19

Ferrous perchlorate is $Fe(ClO_4)_2$ and stannic chromate is $Sn(CrO_4)_2$.

Exercise 7.20

Ammonium nitrate is NH_4NO_3.

Exercise 7.21
H_2CO_3 contains hydrogen and carbonate ions, thus it is carbonic acid. H_3PO_4 is phosphoric acid.

Exercise 7.22
Hypochlorous acid is HClO. ClO^- combines with one H^+.

Exercise 7.23
NH_4HS is ammonium hydrogen sulfide or ammonium bisulfide. KRb_2PO_4 is potassium dirubidium phosphate.

Exercise 7.24
Sodium bicarbonate contains Na^+ and HCO_3^- and is thus $NaHCO_3$.

C. Chapter 7 Practice Test

1. In assigning oxidation numbers to atoms,
   a. halogens are always assigned -1.
   b. elements are assigned zero.
   c. sodium sometimes is assigned +2.
   d. hydrogen is usually +1.

2. Assign oxidation numbers to each atom in $Cr_2O_7^{2-}$ and $Ca(MnO_4)_2$.

3. Binary compounds are comprised of: __________ and _________ .

4. Binary compounds are named
   a. ending in -ide.
   b. always using Greek prefixes.
   c. with oxidation numbers in parentheses after each element.
   d. with metals preceding nonmetals.

5. Name the following compounds using both the Stock and the old names if appropriate: CrI_3, Li_2S, Zn_3P_2, VO_2.

6. Name each of the following, N_2O, SF_6, P_2O_5.

7. The compound Cd_3As_2 contains __________ cations and ________ anions.

8. Write formulas for these compounds:
   tellurium disulfide
   silicon carbide
   sodium carbonate
   tin(II) sulfite
   copper(II) phosphate

9. Lithium potassium carbonate contains the ions:
   a. Li^+
   b. Li^+, K^+, C^{4-}, O^{2-}

c. LiK^{2+}, CO_3^{2-}

d. Li^+, K^+, CO_3^{2-}

10. Name the following aqueous acids or write their formulas, HBr, sulfurous, $HMnO_4$, hypobromous.

D. Answers To Practice Test:

1. Only b and d are true.

2. $Cr_2O_7^{2-}$ Cr +6, O -2

   $Ca(MnO_4)_2$: Ca +2, Mn +7, O -2

3. metals and nonmetals

4. Only a and d are true.

5.

	Stock	old
CrI_3	chromium(III) iodide	chromic iodide
Li_2S	lithium sulfide	
Zn_3P_2	zinc phosphide	
VO_2	vanadium(IV) oxide	-none-

6. dinitrogen monoxide, sulfur hexafluoride, diphosphorus pentoxide

7. Cd^{2+}, As^{3-}

8. TeS_2, SiC, Na_2CO_3, $SnSO_3$, $Cu_3(PO_4)_2$

9. d

10. hydrobromic acid, H_2SO_3, permanganic acid, BrOH

Chapter 8

Atomic Weight, Molecular Weight, The Mole

A. Outline and Study Hints

Section 8.1

- Each isotope of a particular element has a different mass number given as a left hand superscript. ^{12}C and ^{13}C both have 6 protons (the atomic number of carbon is 6) but ^{12}C has 12 - 6 = 6 neutrons whereas ^{13}C has 13 - 6 = 7 neutrons. It has been internationally agreed to assign ^{12}C an arbitrary relative mass of exactly 12 amu, where one amu = 1.6606×10^{-24} g. Masses of all other isotopes are approximately equal to their mass number or more exactly, measured using a mass spectrometer.

Section 8.2

- Because natural samples of elements are mixtures of isotopes (isotopic mixtures), average values of isotopic masses are used for atomic weights. The atomic weight is obtained by summing the products of isotopic abundances and masses, then dividing by 100.

Section 8.3

- By assigning an arbitrary atomic mass to one element, the atomic masses of other elements can be obtained by weighing reactants and products. This is how atomic masses were deduced before mass spectrometers were invented in the middle 1900's.

Section 8.4

- The mole is used to count atoms in the same way that dozens are used to count eggs and millions are used to count dollars (if you're wealthy!). The mole, abbreviated mol, is equal to Avogadro's number, $N_A = 6.022 \times 10^{23}$. We use the conversion factor

$$\frac{6.022 \times 10^{23} \text{ objects}}{\text{mol}} \text{ or } \frac{\text{mol}}{6.022 \times 10^{23} \text{ objects}}$$

where objects could be atoms, molecules, formula units or mangos (a tropical fruit).

Section 8.5

- The atomic mass in grams of any element contains Avogadro's number of atoms. Alternatively we say that the atomic weight in grams is one mole. This provides numerous conversion factors, such as

for aluminum: $$\frac{26.98 \text{ g Al}}{\text{mol Al}} \text{ or } \frac{\text{mol Al}}{26.98 \text{ g Al}}$$

or lithium: $$\frac{6.94 \text{ g Li}}{6.022 \times 10^{23} \text{ atoms Li}} \text{ or } \frac{6.022 \times 10^{23} \text{ atoms Li}}{6.94 \text{ g Li}}$$

- The number of grams per mole of a substance is called the molar mass of that substance. The molar mass of an element is its atomic weight in units of grams per mole.

Section 8.6

- Molecular weight, also called molecular mass or formula mass, is calculated by summing the atomic weights of the elements in the formula of the compound.

Section 8.7

- One mole of a compound contains Avogadro's number of formula units (or molecules) and has a mass equal to the formula weight in grams. This mass, having units of g/mol, is called the molar mass. Note that this definition includes the specialized case for elements discussed in Section 8.5. The precise definition should be committed to memory. Perhaps you can chant it like a mantra!
- Conversion factors are utilized in calculating masses, molecules (or formula units), and moles. The solution map is summarized below where, for example, to convert from grams to moles, you would divide the mass by the molar mass.

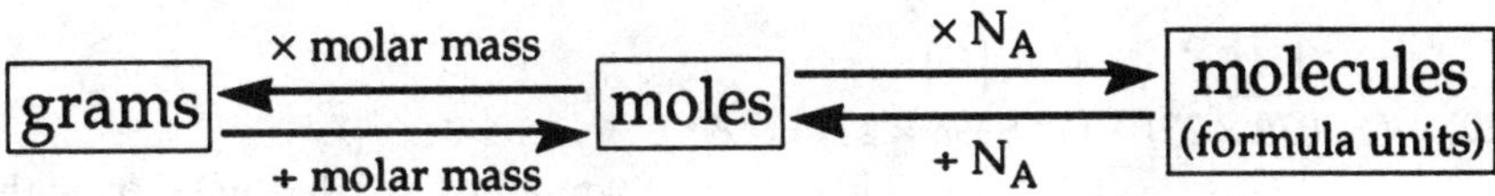

- You should study the exercises and examples in the text before attempting problems. You would do well to "think moles" in your approach to chemical problems.

B. Solutions to In-Text Exercises

Exercise 8.1

$78.9\ \text{amu/atom} \times 1.6606 \times 10^{-24}\ \text{g/amu} = 1.31 \times 10^{-22}\ \text{g/atom}\ ^{79}\text{Br}$

Exercise 8.2

$100.00\ \% - 50.54\ \%\ ^{79}\text{Br} = 49.46\ \%\ ^{81}\text{Br}$ (sum of abundances must equal 100 %)

$$\frac{(50.54 \times 78.9183) + (49.46 \times 80.9163)}{100} = 79.91, \text{ atomic weight of Br.}$$

Exercise 8.3

$$4.92 \times 10^{24}\ \text{atoms Cu} \times \frac{\text{mol}}{6.022 \times 10^{23}\ \text{atoms}} = 8.17\ \text{mol of Cu}$$

Exercise 8.4

$$22.7\ \text{mol H}_2\text{O} \times \frac{6.022 \times 10^{23}\ \text{H}_2\text{O molecules}}{\text{mol H}_2\text{O}} = 1.37 \times 10^{25}\ \text{H}_2\text{O molecules}$$

Exercise 8.5

$$0.0562\ \text{mol Ag} \times \frac{107.868\ \text{g Ag}}{\text{mol Ag}} = 6.06\ \text{g Ag}$$ (final answer rounded to 3 significant figures).

Exercise 8.6

$$32.6\ \text{g Cu} \times \frac{\text{mol Cu}}{63.546\ \text{g Cu}} = 0.513\ \text{mol Cu}$$

Exercise 8.7

$$1.5\text{ g} \times \frac{\text{mol Na}}{22.9898\text{g Na}} \times \frac{6.022 \times 10^{23}\text{ Na atoms}}{\text{mol Na}} = 3.9 \times 10^{22}\text{ Na atoms}$$

(rounded to 2 significant figures)

Exercise 8.8

$$0.729\text{ g Ca} \times \frac{\text{mol Ca}}{40.08\text{ g Ca}} \times \frac{6.022 \times 10^{23}\text{ Ca atoms}}{\text{mol Ca}} = 1.10 \times 10^{22}\text{ Ca atoms}$$

Exercise 8.9

$$4.62 \times 10^{24}\text{ atoms Al} \times \frac{\text{mol Al}}{6.022 \times 10^{23}\text{ Al atoms}} \times \frac{26.9815\text{ g Al}}{\text{mol Al}} = 207\text{ g Al}$$

Exercise 8.10

$$\boxed{\text{C}} \qquad \boxed{\text{H}} \qquad \boxed{\text{O}}$$

$$(12.011 \times 12) + (1.0079 \times 22) + (15.9994 \times 11) =$$

$$(144.132) + (22.1738) + (175.9934) = 342.299 \left(\begin{array}{c}\text{rounded to 3 decimal}\\ \text{places because 12.011}\\ \text{has 3 decimal places}\end{array}\right)\left(\begin{array}{c}\text{note this result}\\ \text{has 6 significant}\\ \text{figures}\end{array}\right)$$

Exercise 8.11

$$\boxed{\text{Ba}} \qquad \boxed{\text{F}}$$

$$137.327 + 2(18.9984) = 175.323$$

$$\boxed{\text{Fe}} \qquad \boxed{\text{Cr}} \qquad \boxed{\text{O}}$$

$$2(55.847) + 6(51.9961) + 21(15.9994) = 759.66$$

$$\boxed{\text{K}} \qquad \boxed{\text{Br}} \qquad \boxed{\text{O}}$$

$$39.0983 + 79.904 + 3(15.9994) = 167.000$$

Exercise 8.12

$$\boxed{\text{N}} \qquad \boxed{\text{O}}$$

$$14.0067 + 2(15.9994) = 46.0055\text{ g/mol}$$

$$\boxed{\text{B}} \qquad \boxed{\text{F}}$$

$$10.811 + 3(18.9984) = 67.806\text{ g/mol}$$

Molar mass has the units of g/mol. The molecular weight is in amu or without units.

Exercise 8.13

$$4.8\text{ mol } I_2 \times \frac{253.809\text{ g } I_2}{\text{mol } I_2} = 1.2 \times 10^3\text{ g } I_2$$

Exercise 8.14

$$45.32\text{ g } K_2Cr_2O_7 \times \frac{\text{mol } K_2Cr_2O_7}{294.18\text{ g } K_2Cr_2O_7} = 0.1541\text{ mol } K_2Cr_2O_7$$

Exercise 8.15

We could first calculate the number of moles of C_2H_6

$$34.8 \text{ g } C_2H_6 \times \frac{\text{mol } C_2H_6}{30.069 \text{ g } C_2H_6} = 1.157 \text{ mol } C_2H_6$$

and then calculate the molecules

$$1.157 \text{ mol } C_2H_6 \times \frac{6.022 \times 10^{23} \text{ } C_2H_6 \text{ molecules}}{\text{mol } C_2H_6} = 6.97 \times 10^{23} \text{ } C_2H_6 \text{ molecules}$$

Alternatively, we could write both conversions on the same line,

$$34.8 \text{ g } C_2H_6 \times \frac{\text{mol } C_2H_6}{30.069 \text{g } C_2H_6} \times \frac{6.022 \times 10^{23} \text{ } C_2H_6 \text{ molecules}}{\text{mol } C_2H_6} = 6.97 \times 10^{23} \text{ } C_2H_6 \text{ molecules}$$

Exercise 8.16

$$6.97 \times 10^{23} \text{ } N_2 \text{ molecules} \times \frac{\text{mol } N_2}{6.022 \times 10^{23} \text{ } N_2 \text{ molecules}} \times \frac{28.0134 \text{ g } N_2}{\text{mol } N_2} = 32.4 \text{ g } N_2$$

Exercise 8.17

$$0.491 \text{ mol KBr} \times \frac{119.002 \text{ g KBr}}{\text{mol KBr}} = 58.4 \text{ g KBr}$$

Exercise 8.18

$$9.0 \text{ g KI} \times \frac{\text{mol KI}}{166.003 \text{ g KI}} = 0.054 \text{ mol KI}$$

C. Chapter 8 Practice Test

1. Atomic weight of an element is obtained by averaging the ______ of the isotopes taking into account their ______.

2. Isotopes of a given element
   a. have equal numbers of protons and electrons.
   b. have the same mass.
   c. differ in their mass number.
   d. differ in the number of protons.

3. Boron consists of two isotopes ^{10}B and ^{11}B. The atomic weight of boron is 10.8. Which isotope is more abundant?

4. A grain of rice weighs about 10 mg. How many kilograms would a mole of rice grains weigh?

5. Six moles of helium
   a. weighs 4 g.
   b. contains 6.022×10^{23} atoms.
   c. weighs 24 g.
   d. contains 36×10^{23} atoms.

6. Calculate the mass in grams of a million sodium atoms.

7. Find the formula weight of CCl_4, HIO_4, and $Ca(MnO_4)_2$.

8. Which statement(s) are true?
   a. Molar mass has amu units.
   b. Molecular weight and molar mass have the same value but different units.
   c. Molecular weight for elements is atomic weight.
   d. Molecular weight in grams contains Avogadro's number of formula units.

9. Calculate the number of moles of carbon tetrachloride in 100 g of CCl_4.

10. Calculate the mass in grams of 6.022×10^{10} $Ca(MnO_4)_2$ formula units.

D. Answers To Practice Test:

1. atomic masses, isotopic abundances

2. a and c are true

3. ^{11}B, because the atomic weight is closer to 11 than 10.

4. 6×10^{18} kg!

5. c and d

6. 3.8×10^{-17} g Na

7. 153.82, 191.91, 277.95

8. b, c, d

9. 0.65 mol CCl_4

10. 2.779×10^{-11} g

Chapter 9

Mass Relations in Compounds

A. Outline and Study Hints

Section 9.1

- You should have memorized the solution map,

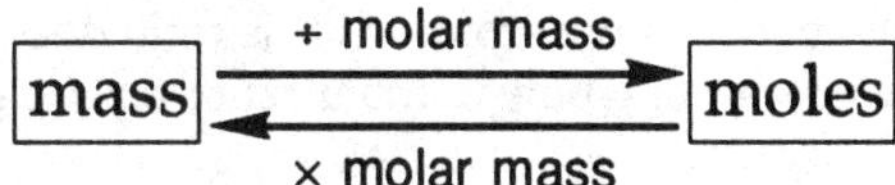

If you do not get Exercises 9.1 through 9.4 correct, you need to do more problems from chapter 8 before you continue chapter 9.

Section 9.2

- The molecular formula tells us the number and type of atoms. The formula of chlorous acid, $HClO_2$, allows us to construct the conversion factors

$$\frac{\text{mol H}}{\text{mol HClO}_2}, \frac{\text{mol Cl}}{\text{mol HClO}_2}, \frac{\text{2 mol O}}{\text{mol HClO}_2} \text{ as well as their reciprocals.}$$

- Complex calculations involving mass and moles of a particular element may now be done with these new conversion factors. Do not go on to section 9.3 until you have studied the text examples and you have done the exercises correctly.

Section 9.3

- The percent composition of a compound is the number of grams of each element in 100 grams of compound. It is convenient to calculate the molar mass of the compound and the mass due to each element in performing percent composition calculations. Percent (or percentage) is defined as

$$\% = \left(\frac{\text{part}}{\text{whole}}\right) \times 100\ \%,$$ where the units of "part" and "whole" must be the same.

The ratio $\left(\frac{\text{part}}{\text{whole}}\right)$ is called the fraction.

- The sum of all percentages must equal 100%

Section 9.4

- Following the prescription in this section, you can calculate the percent composition from the formula:

$$\frac{\text{mol element}}{\text{mol compound}} \times \frac{\text{g element}}{\text{mol element}} \times \frac{\text{mol compound}}{\text{g compound}} \times 100\% = \%\ \text{element}$$

Section 9.5

- An empirical formula is the simplest formula whereas a molecular represents the actual formula. Several molecules have the empirical formula CH_2O, two whose Lewis diagrams are shown.

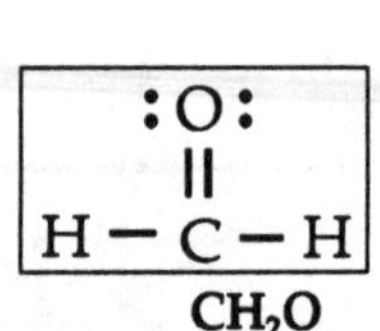

CH_2O

```
H :O:
|  ||  ..
H − C − C − O − H
|        ..
H
```

$C_2H_4O_2$

- Empirical formulas can be obtained from percent composition or from mass data. First find the number of moles of each element directly from the mass or percent. Then divide these by the smallest; this results in a ratio between the atoms. If this ratio does not consist of whole numbers, then multiply by a whole number until whole number ratios are found.

Section 9.6

- Molecular formulas can be calculated from empirical formulas if the molecular weight is known. Dividing the molecular weight by the empirical formula weight yields the number by which all the subscripts in the empirical formula must be multiplied to obtain the molecular formula. Note that the order of atoms in a formula is not crucial, thus CH_2O could also be written COH_2.

B. Solutions to In-Text Exercises

Exercise 9.1

$$10.5 \text{ g He} \times \frac{\text{mol He}}{4.0026 \text{ g He}} = 2.62 \text{ mol He}$$

Exercise 9.2

$$0.0562 \text{ mol Ag} \times \frac{107.8682 \text{ g Ag}}{\text{mol Ag}} = 6.06 \text{ g Ag}$$

Exercise 9.3

$$176.2 \text{ g } K_2SO_4 \times \frac{\text{mol } K_2SO_4}{174.260 \text{ g } K_2SO_4} = 1.011 \text{ mol } K_2SO_4$$

Exercise 9.4

$$1.55 \text{ mol CO} \times \frac{28.010 \text{ g CO}}{\text{mol CO}} = 43.4 \text{ g CO}$$

Exercise 9.5

$$1.31 \text{ mol } C_6H_5O_4N \times \frac{4 \text{ mol O}}{\text{mol } C_6H_5O_4N} = 5.24 \text{ mol O}$$

Exercise 9.6

$$4.6 \text{ mol H} \times \frac{\text{mol } H_2O}{2 \text{ mol H}} = 2.3 \text{ mol } H_2O$$

Exercise 9.7

The solution map is $\boxed{g\ Al_2O_3} \rightarrow \boxed{mol\ Al_2O_3} \rightarrow \boxed{mol\ O}$

Converting first to mol Al_2O_3 and then to mol O yields,

$$320.5\ g\ Al_2O_3 \times \frac{mol\ Al_2O_3}{101.9612\ g\ Al_2O_3} \times \frac{3\ mol\ O}{mol\ Al_2O_3} = 9.430\ mol\ O$$

Exercise 9.8

The solution map is $\boxed{g\ CO_2} \rightarrow \boxed{mol\ CO_2} \rightarrow \boxed{mol\ O}$

$$156\ g\ CO_2 \times \frac{mol\ CO_2}{44.010\ g\ CO_2} \times \frac{2\ mol\ O}{mol\ CO_2} = 7.09\ mol\ O$$

Exercise 9.9

The solution map is $\boxed{g\ Cr_2(SO_4)_3} \rightarrow \boxed{mol\ Cr_2(SO_4)_3} \rightarrow \boxed{mol\ O} \rightarrow \boxed{g\ O}$

$$82\ g\ Cr_2(SO_4)_3 \times \frac{mol\ Cr_2(SO_4)_3}{392.183\ g\ Cr_2(SO_4)_3} \times \frac{12\ mol\ O}{mol\ Cr_2(SO_4)_3} \times \frac{15.9994\ g\ O}{mol\ O} = 40.\ g\ O$$

Exercise 9.10

The solution map is $\boxed{g\ H_2S} \rightarrow \boxed{mol\ H_2S} \rightarrow \boxed{mol\ H} \rightarrow \boxed{g\ H}$

$$46.3\ g\ H_2S \times \frac{mol\ H_2S}{34.082\ g\ H_2S} \times \frac{2\ mol\ H}{mol\ H_2S} \times \frac{1.0079\ g\ H}{mol\ H} = 2.74\ g\ H$$

Exercise 9.11

Dividing the oxygen mass by mass of the whole compound

$$\frac{3.20\ g}{4.40\ g} \times 100\% = 72.7\ \%\ oxygen$$

Exercise 9.12

Dividing the part that is nitrogen by the whole

$$\frac{0.70\ kg\ N}{2.00\ kg\ sample} \times 100\% = 35\%\ nitrogen$$

Exercise 9.13

The total mass is the sum of all the masses

$$1.20\ g + 0.10\ g + 0.70\ g = 2.00\ g$$

$$\frac{1.20\ g}{2.00\ g} \times 100\% = 60.0\%\ C; \quad \frac{0.10\ g}{2.00\ g} \times 100\% = 5.0\%\ H; \quad \frac{0.70\ g}{2.00\ g} \times 100\% = 35\%\ N$$

(note the sum of the percentages 60.0 + 5.0 + 35 = 100)

Exercise 9.14

$0.403\text{ g H} + 6.412\text{ g S} + 12.800\text{ g O} = 19.615\text{ g acid}$

$$\frac{0.403\text{ g H}}{19.615\text{ g acid}} \times 100\% = 2.05\%\text{ H}; \quad \frac{6.412\text{ g S}}{19.615\text{ g acid}} \times 100\% = 32.69\%\text{ S};$$

$$\frac{12.800\text{ g O}}{19.615\text{ g acid}} \times 100\% = 65.256\%\text{ O}$$

Exercise 9.15

The mass of Br is obtained by subtracting other masses from the weight of the sample.

$118.9\text{ g sample} - (23.0\text{ g Na} + 16.0\text{ g O}) = 79.9\text{ g Br}$

$$\frac{23.0\text{ g Na}}{118.9\text{ g sample}} \times 100\% = 19.3\%\text{ Na}; \frac{16.0\text{ g O}}{118.9\text{ g sample}} \times 100\% = 13.5\%\text{ O}$$

$$\frac{79.9\text{ g Br}}{118.9\text{ g sample}} \times 100\% = 67.2\%\text{ Br}$$

Exercise 9.16

Each mol of NaOH contains 1 mol each of Na, O and H. The formula mass of NaOH is 39.9971.

$$\frac{\text{mol Na}}{\text{mol NaOH}} \times \frac{22.9898\text{ g Na}}{\text{mol Na}} \times \frac{\text{mol NaOH}}{39.9971\text{ g NaOH}} \times 100\% = 57.4787\%\text{ Na}$$

$$\frac{\text{mol O}}{\text{mol NaOH}} \times \frac{15.9994\text{ g O}}{\text{mol O}} \times \frac{\text{mol NaOH}}{39.9971\text{ g NaOH}} \times 100\% = 40.0014\ \%\text{ O}$$

$$\frac{\text{mol H}}{\text{mol NaOH}} \times \frac{1.0079\text{ g H}}{\text{mol H}} \times \frac{\text{mol NaOH}}{39.9971\text{ g NaOH}} \times 100\% = 2.5199\ \%\text{ H}$$

Exercise 9.17

One mole of $Ca(ClO_2)_2$ contains two mol Cl, four mol O and one mol Ca. The formula mass of $Ca(ClO_2)_2$ is 174.981.

$$\frac{\text{mol Ca}}{\text{mol Ca(ClO}_2)_2} \times \frac{40.08\text{ g Ca}}{\text{mol Ca}} \times \frac{\text{mol Ca(ClO}_2)_2}{174.981\text{ g Ca(ClO}_2)_2} \times 100\% = 22.905\%\text{ Ca}$$

$$\frac{2\text{ mol Cl}}{\text{mol Ca(ClO}_2)_2} \times \frac{35.453\text{ g Cl}}{\text{mol Cl}} \times \frac{\text{mol Ca(ClO}_2)_2}{174.981\text{ g Ca(ClO}_2)_2} \times 100\% = 40.5221\ \%\text{ Cl}$$

$$\frac{4\text{ mol O}}{\text{mol Ca(ClO}_2)_2} \times \frac{15.9994\text{ g O}}{\text{mol O}} \times \frac{\text{mol Ca(ClO}_2)_2}{174.981\text{ g Ca(ClO}_2)_2} \times 100\% = 36.5740\%\text{ O}$$

Exercise 9.18

Finding the mol of each element

$$3.0\,\text{g C} \times \frac{\text{mol C}}{12.011\,\text{g C}} = 0.25\,\text{mol C}$$

$$1.0\,\text{g H} \times \frac{\text{mol H}}{1.0079\,\text{g H}} = 0.99\,\text{mol H};\ \ 4.0\,\text{g O} \times \frac{\text{mol O}}{15.9994\,\text{g O}} = 0.25\,\text{mol O}$$

then divide each by the smallest (0.25) and round to 2 significant figures.

$$\frac{0.25\,\text{mol C}}{0.25} = 1.0\,\text{mol C};\ \ \frac{0.99\,\text{mol H}}{0.25} = 4.0\,\text{mol H};\ \ \frac{0.25\,\text{mol O}}{0.25} = 1.0\,\text{mol O}.$$

Thus $C_1H_4O_1$ or CH_4O is the empirical formula.

Exercise 9.19

Computing the number of mol of each element

$$5.06\,\text{g Na} \times \frac{\text{mol Na}}{22.9898\,\text{g Na}} = 0.220\,\text{mol Na}$$

$$2.64\,\text{g C} \times \frac{\text{mol C}}{12.011\,\text{g C}} = 0.220\,\text{mol C}$$

$$10.56\,\text{g O} \times \frac{\text{mol O}}{15.9994\,\text{g O}} = 0.6600\,\text{mol O}$$

$$0.222\,\text{g H} \times \frac{\text{mol H}}{1.0079\,\text{g H}} = 0.220\,\text{mol H}$$

and dividing by the smallest (0.220)

$$\frac{0.220\,\text{mol Na}}{0.220} = 1.00\,\text{mol Na};\ \ \frac{0.220\,\text{mol C}}{0.220} = 1.00\,\text{mol C};$$

$$\frac{0.6600\,\text{mol O}}{0.220} = 3.00\,\text{mol O};\ \ \frac{0.220\,\text{mol H}}{0.220} = 1.00\,\text{mol H};$$

yields the empirical formula $NaHCO_3$.

Exercise 9.20

Using the percentages as mass per 100 g of compound,

$$52.0\text{ g Zn} \times \frac{\text{mol Zn}}{65.39\text{ g Zn}} = 0.795\text{ mol Zn}$$

$$9.61\text{ g C} \times \frac{\text{mol C}}{12.011\text{ g C}} = 0.800\text{ mol C}$$

$$38.4\text{ g O} \times \frac{\text{mol O}}{15.9994\text{ g O}} = 2.40\text{ mol O}$$

then dividing by the smallest (0.795) and rounding to two significant figures.

$$\frac{0.795\text{ mol Zn}}{0.795} = 1.00\text{ mol Zn;}$$

$$\frac{0.800\text{ mol C}}{0.795} = 1.01\text{ mol C;}$$

$$\frac{2.40\text{ mol O}}{0.795} = 3.02\text{ mol O;}$$

yields the empirical formula $ZnCO_3$.

Exercise 9.21

Calculating the number of moles of C and H

$$11.25\text{ g C} \times \frac{\text{mol C}}{12.011\text{ g C}} = 0.9366\text{ mol C; } 2.52\text{ g H} \times \frac{\text{mol H}}{1.0079\text{ g H}} = 2.50\text{ mol H}$$

dividing by the smallest (0.9366)

$$\frac{0.9366\text{ mol C}}{0.9366} = 1.000\text{ mol C}; \quad \frac{2.50\text{ mol H}}{0.9366} = 2.67\text{ mol H}$$

and multiplying by 3 yields whole numbers.

3×1.000 mol C = 3.000 mol C; 3×2.67 mol H = 8.01 mol H; thus the empirical formula is C_3H_8.

Exercise 9.22

Once again, using the percentages as mass per 100 g compound,

$$72.36\text{ g Fe} \times \frac{\text{mol Fe}}{55.847\text{ g Fe}} = 1.296\text{ mol Fe; } 27.64\text{ g O} \times \frac{\text{mol O}}{15.9994\text{ g O}} = 1.728\text{ mol O}$$

$$\frac{1.296\text{ mol Fe}}{1.296} = 1.000\text{ mol Fe}; \quad \frac{1.728\text{ mol O}}{1.296} = 1.333\text{ mol O}$$

3×1.000 mol Fe = 3.000 mol Fe; 3×1.333 mol O = 3.999 mol O.

This is essentially 3 to 4, thus the empirical formula is Fe_3O_4.

Exercise 9.23

The empirical weight is 12 + 3 + 16 = 31, thus $\frac{60}{31} \approx 2$ implying a formula $C_2H_6O_2$ whose molecular weight is

$$2(12.011) + 6(1.0079) + 2(15.9994) = 62.068 \text{ g/mol}$$

Exercise 9.24

The empirical formula mass of C_4H_3OCl is about 102

$$\frac{200}{102} \approx 2 \text{ thus } C_8H_6O_2Cl_2$$

$$8(12.011) + 6(1.0079) + 2(15.9994) + 2(35.453) = 205.040 \text{ g/mol}$$

Exercise 9.25

First find the empirical formula.

$$37.6 \text{ g C} \times \frac{\text{mol C}}{12.0 \text{ g C}} = 3.13 \text{ mol C}$$

$$4.20 \text{ g H} \times \frac{\text{mol H}}{1.008 \text{ g H}} = 4.17 \text{ mol H}$$

$$58.37 \text{ g O} \times \frac{\text{mol O}}{16.00 \text{ g O}} = 3.648 \text{ mol O}$$

$$\frac{3.13 \text{ mol C}}{3.13} = 1.00 \text{ mol C}; \quad \frac{4.17 \text{ mol H}}{3.13} = 1.33 \text{ mol H}; \quad \frac{3.648 \text{ mol O}}{3.13} = 1.17 \text{ mol O}$$

Multiplication by 6 yields whole numbers.

$$6 \times 1.00 \text{ mol C} = 6.00 \text{ mol C}; \; 6 \times 1.33 \text{ mol H} = 7.98 \text{ mol H}; \; 6 \times 1.17 \text{ mol O} = 7.02 \text{ mol O},$$

Thus, the empirical formula is $C_6H_8O_7$, whose formula weight is

$$6(12.011) + 8(1.0079) + 7(15.994) = 192.125;$$

which is approximately the same as the given molecular weight of 190. The empirical formula in this case is also the molecular formula.

C. Chapter 9 Practice Test

1. In converting mass to moles, you divide the _______ by the _______.

2. A molecular formula states the number of _______ of each element that are contained per _______. These numbers can also be interpreted as the number of _______ of each element per _______ of compound.

3. How many moles of oxygen atoms are contained in six moles of dinitrogen pentoxide?
   a. one
   b. six
   c. five
   d. thirty

4. The number of grams of oxygen in six moles of dinitrogen pentoxide is
   a. 16.
   b. 6×16.
   c. 5×16.
   d. 30×16.

5. What is the percent composition in a compound formed by combining 222.6 g of sodium with 77.4 g of oxygen?

6. What mass of carbon is contained in 82 g of ethane, C_2H_6?
   a. 82 g
   b. 100 g
   c. 66 g
   d. 8 g

7. Write the formula for calcium acetate and calculate its percent composition.

8. Calculate the empirical formula of a compound that is 25.9 % nitrogen and 74.1% oxygen.

9. A compound is formed by combining 9.03 g of Mg with 3.48 g of N. Deduce the empirical formula from this data.

10. An ester that smells like apples has been found to have a molecular weight of 102 and a percent composition of 58.8 % C, 9.8% H and 31.4% O. What is its molecular formula?

D. Answers To Practice Test:

1. mass (in grams), molar mass

2. atoms, molecule, moles, mole

3. d

4. d

5. 74.2 % Na, 25.8 % O

6. c

7. $Ca(C_2H_3O_2)_2$; 25.339% Ca, 30.375 % C, 3.8234% H, 40.4620 % O

8. N_2O_5

9. Mg_3N_2

10. $C_5H_{10}O_2$

Chapter 10

Calculations Using Information from Chemical Equations

A. Outline and Study Hints

Section 10.1

- A chemical equation represents a chemical reaction. The left hand side of the equation is called the reactant side where formulas represent the reactants separated by plus signs. The right hand side contains formulas of products also separated by plus signs, just as in algebra. The states (g, l, s) are shown in parentheses after each reactant and product. An arrow (→) separates reactants and products. Sometimes other symbols are used: =, $\leftrightarrows$ or variants of these. Reactants and products must be known before writing a chemical equation.

Section 10.2

- A chemical equation must be balanced; the number of atoms of each element type must be equal on the reactant and product sides. This is done by varying the coefficients in front of the formulas. No one set of coefficients is correct, but we often use the smallest integer coefficients. Balancing is done by trial and error with lots of practice. It is helpful to set up a table to check whether each atom is balanced on both sides. After a lot of practice (about 50 equations!), you can then do this in your head. You should practice on speed as well as accuracy.
An example is exercise 10.2:

	$Zn(s)$ +	$NH_4NO_3(s)$ →	$N_2(g)$ +	$ZnO(s)$ +	$H_2O(g)$
Zn:	1			1	
N:		2	2		
H:		4			2
O:		3		1	1

The table shows that Zn and N are presently balanced but H and O are not. We need 2 H_2O on the right hand side which will increase H to 4 and O to 3 on the right hand side, thus the balanced equation is

$$Zn(s) + NH_4NO_3(s) \longrightarrow N_2(g) + ZnO(s) + 2\ H_2O(g)$$

zinc	ammonium nitrate	nitrogen	zinc oxide	water

- The names are placed below the compounds as a reminder that you should practice your naming rules throughout this chapter.

Section 10.3

- Chemical equations coefficients can be read in units of molecules (or formula units for ionic compounds) or moles. The equation above could be interpreted to mean that if one mole of zinc reacts with one mole of ammonium nitrate, then one mole of nitrogen, one mole of zinc oxide and two moles of water will be formed.

Section 10.4

- Conversion factors that relate moles of reactants and products are derived from balanced chemical equations. The fact that in the above reaction, one mole of Zn produces two moles of H_2O can be displayed as conversion factors:

$$\frac{\text{mol Zn}}{\text{2 mole } H_2O} \text{ or } \frac{\text{2 mol } H_2O}{\text{mole Zn}}$$

Section 10.5

- Stoichiometric calculations can be performed using conversion factors. The quantity desired is first placed on the right hand side. For example, in exercise 10.4, the moles of H_2 is being sought, thus

$$\boxed{} = \text{? mol } H_2$$

and the starting point is to write the amount given on the left hand side:

$$\text{5.45 mol NaOH} \times \left(\frac{}{}\right) = \text{? mol } H_2$$

The parentheses encloses the conversion factor obtained from the balanced equation.

- The units must be mol NaOH in the denominator and mol H_2 in the numerator to yield the mol H_2 shown on the right hand side:

$$\text{5.45 mol NaOH} \times \left(\frac{\text{? mol } H_2}{\text{? mol NaOH}}\right) = \text{? mol } H_2$$

- Now the question marks can be filled in from the balanced equation which shows that 1 mol of H_2 is formed from 2 mol of NaOH:

$$\text{5.45 mol NaOH} \times \frac{\text{mol } H_2}{\text{2 mol NaOH}} = \text{2.73 mol } H_2$$

Section 10.6

- We can perform stoichiometric calculations using masses by first converting to moles. Thus the solution map is:

$$\boxed{\text{grams A}} \xrightarrow[\text{of A}]{\text{I: } \div \text{ molar mass}} \boxed{\text{moles A}} \xrightarrow[\text{from balanced equation}]{\text{II: stoichiometric conversion}} \boxed{\text{moles B}} \xrightarrow[\text{of B}]{\text{III: } \times \text{ molar mass}} \boxed{\text{grams B}}$$

- The three steps (I, II, III) can be done in one calculation involving three conversion factors. You must always start with a balanced chemical equation when performing these calculations. It is usually best to write the units of the conversion factors first, and then fill in the numerical values. Let's calculate the mass of Zn needed to produce 1950 grams of $Zn(NO_3)_2$, thus

	I		II	III	
g $Zn(NO_3)_2$	→	mol $Zn(NO_3)_2$	→	mol Zn	→ g Zn

$$1950 \text{ g } Zn(NO_3)_2 \times \frac{1 \text{ mol } Zn(NO_3)_2}{189.3998 \text{ g } Zn(NO_3)_2} \times \frac{\text{mol Zn}}{\text{mol } Zn(NO_3)_2} \times \frac{65.39 \text{ g Zn}}{\text{mol Zn}} = ?\text{ g Zn or } 673 \text{ g Zn}$$

↑ molar mass of $Zn(NO_3)_2$ ↑ stoichiometry ↑ molar mass of Zn

Section 10.7

- The actual yield is the amount really produced in a chemical reaction. This amount is almost always less than the theoretical yield that we have been calculating throughout this chapter. The percentage yield or percent yield is given by

$$\frac{\text{actual yield}}{\text{theoretical yield}} \times 100\%$$

Section 10.8

- When several reactants are mixed, often one of them acts as the limiting reactant or reagent and is thus completely used up or consumed in the process. The reactant that theoretically produces the least amount of product will be the limiting reactant. Other reactants will be in excess.

Section 10.9

- In mass calculations, remember that you must first calculate moles before further calculations can be done. The solution map from section 10.6 above should be kept in mind when you work problems. Think moles!

Section 10.10

- Density data can be used to convert from volume to mass data. The density of mercury, 13.5 g/mL, allows us to write two conversion factors:

$$\frac{13.5 \text{ g Hg}}{\text{mL Hg}} \text{ and } \frac{\text{mL Hg}}{13.5\text{g Hg}}$$

The solution map

volume —(× density)→ mass
volume ←(÷ density)— mass

can be combined at the ends of the solution map from section 10.6 to do stoichiometric calculations.

Section 10.11

- Energy in the form of heat is often either consumed (endothermic, +ΔH) or produced (exothermic, -ΔH) during a chemical reaction. Combustion (reacting with oxygen to produce CO_2 and H_2O) reactions are exothermic.

B. Solutions to In-Text Exercises.

Exercise 10.1

	$NaCl$ +	$3\ NaBrO \rightarrow$	$NaClO_3$ +	$3\ NaBr$.
Na:	1	3	1	3
Cl:	1	0	1	0
Br:	0	3	0	3
O:	0	3	3	0

Exercise 10.2

$Zn(s) + NH_4NO_3(s) \longrightarrow N_2(g) + ZnO(s) + 2\ H_2O(g)$.

Exercise 10.3

Using the conversion that one mol H_2 is required for each mol of CO,

$$1.60 \text{ mol CO} \times \frac{\text{mol } H_2}{\text{mol CO}} = 1.60 \text{ mol } H_2$$

Exercise 10.4

2 mol of NaOH are required to produce 1 mol H_2, thus

$$5.45 \text{ mol NaOH} \times \frac{\text{mol } H_2}{2 \text{ mol NaOH}} = 2.732 \text{ mol } H_2 \text{ can be formed.}$$

Exercise 10.5

1 mol of Na_2O is required to produce 2 mol NaCl, thus

$$8.72 \text{ mol NaCl} \times \frac{\text{mol } Na_2O}{2 \text{ mol NaCl}} = 4.36 \text{ mol } Na_2O \text{ are required.}$$

Exercise 10.6

2 mol HCl are required for every 1 mol of H_2O produced, thus

$$3.40 \text{ mol } H_2O \times \frac{2 \text{ mol HCl}}{\text{mol } H_2O} = 6.80 \text{ mol HCl are required.}$$

Exercise 10.7

Noting that 2 mol of C_2H_4 are required to react with 7 mol O_2,

$$3.42 \text{ mol } O_2 \times \frac{2 \text{ mol } C_2H_6}{7 \text{ mol } O_2} = 0.977 \text{ mol } C_2H_6 \text{ are necessary.}$$

Exercise 10.8

Converting first to mol $KClO_3$, then to mol O_2 yields

$$16.4 \text{ g } KClO_3 \times \frac{\text{mol } KClO_3}{122.5492 \text{ g } KClO_3} \times \frac{3 \text{ mol } O_2}{2 \text{ mol } KClO_3} = 0.201 \text{ mol } O_2.$$

Did you remember to balance the equation? $2\ KClO_3 \longrightarrow 2\ KCl + 3\ O_2$.

Exercise 10.9

Using the balanced equation, $2\ HgO \longrightarrow 2\ Hg + O_2$, first calculate mol HgO, then mol Hg.

$$24.3\text{ g HgO} \times \frac{\text{mol HgO}}{216.59\text{ g HgO}} \times \frac{2\text{ mol Hg}}{2\text{ mol HgO}} = 0.112\text{ mol Hg.}$$

Exercise 10.10

Converting to mol Zn followed by gram conversion yields

$$10.3\text{ mol Zn(NO}_3)_2 \times \frac{\text{mol Zn}}{\text{mol Zn(NO}_3)_2} \times \frac{65.39\text{ g Zn}}{\text{mol Zn}} = 674\text{ g Zn.}$$

The equation is balanced as stated in the exercise.

Exercise 10.11

Writing the balanced equation $2\ HgO \longrightarrow 2\ Hg + O_2$, then converting to mol HgO followed by grams HgO yields

$$8.50\text{ mol Hg} \times \frac{2\text{ mol HgO}}{2\text{ mol Hg}} \times \frac{216.59\text{ g HgO}}{\text{mol HgO}} = 1.84 \times 10^3\text{ g HgO.}$$

Exercise 10.12

$$\boxed{\text{g NaOH}} \rightarrow \boxed{\text{mol NaOH}} \rightarrow \boxed{\text{Na}_2\text{ZnO}_2} \rightarrow \boxed{\text{g Na}_2\text{ZnO}_2}$$

$$76.8\text{ g NaOH} \times \frac{\text{mol NaOH}}{39.9971\text{ g NaOH}} \times \frac{\text{mol Na}_2\text{ZnO}_2}{2\text{ mol NaOH}} \times \frac{143.37\text{ g Na}_2\text{ZnO}_2}{\text{mol Na}_2\text{ZnO}_2} = 138\text{ g Na}_2\text{ZnO}_2$$

Exercise 10.13

Find mol Sb, then mol $SbCl_3$ and finally g $SbCl_3$.

$$18.6\text{ g Sb} \times \frac{\text{mol Sb}}{121.75\text{ g Sb}} \times \frac{2\text{ mol SbCl}_3}{2\text{ mol Sb}} \times \frac{228.11\text{ g SbCl}_3}{\text{mol SbCl}_3} = 34.8\text{ g SbCl}_3$$

Exercise 10.14

First calculate the theoretical yield from the 4.6 g HCl

$$4.6\text{ g HCl} \times \frac{\text{mol HCl}}{36.4606\text{ g HCl}} \times \frac{\text{mol NaCl}}{\text{mol HCl}} \times \frac{58.4425\text{ g NaCl}}{\text{mol NaCl}} = 7.4\text{ g NaCl theoretically}$$

then divide the actual (5.7 g) by this and multiply by 100.

$$\frac{5.7\text{ g}}{7.4\text{ g}} \times 100\% = 77\%$$

Exercise 10.15

Calculate the theoretical yield of AgNO3 using mol Ag and then mol HNO3

$$2.5 \text{ mol Ag} \times \frac{3 \text{ mol AgNO}_3}{3 \text{ mol Ag}} = 2.5 \text{ mol AgNO}_3$$

$$2.0 \text{ mol HNO}_3 \times \frac{3 \text{ mol AgNO}_3}{4 \text{ mol HNO}_3} = 1.5 \text{ mol AgNO3}; \text{ thus}$$

HNO_3 is limiting and 1.5 mol $AgNO_3$ will be produced.

Exercise 10.16

Computing maximum mol NO_2 starting with mol Cu and mol HNO_3 yields,

$$0.245 \text{ mol Cu} \times \frac{2 \text{ mol NO}_2}{\text{mol Cu}} = 0.490 \text{ mol NO}_2$$

$$0.320 \text{ mol HNO}_3 \times \frac{2 \text{ mol NO}_2}{4 \text{ mol HNO}_3} = 0.160 \text{ mol NO}_2$$

thus HNO_3 is limiting and 0.160 mol of NO_2 will be produced.

Exercise 10.17

Calculate mol H_3PO_4 from each reactant,

$$0.55 \text{ mol P}_4\,0_{10} \times \frac{4 \text{ mol H}_3\text{PO}_4}{\text{mol P}_4\,0_{10}} = 2.2 \text{ mol H}_3\text{PO}_4$$

$$1.20 \text{ mol H}_2\text{O} \times \frac{4 \text{ mol H}_3\text{PO}_4}{6 \text{ mol H}_2\text{O}} = 0.800 \text{ mol H}_3\text{PO}_4$$

Water is limiting, therefore 0.800 moles of H_3PO_4 will be produced.

Exercise 10.18

Calculating the theoretical yield (in mol) from each reactant,

$$22.6 \text{ g Al} \times \frac{\text{mol Al}}{26.9815 \text{ g Al}} \times \frac{2 \text{ mol AlBr}_3}{2 \text{ mol Al}} = 0.838 \text{ mol AlBr}_3$$

$$38.5 \text{ g Br}_2 \times \frac{\text{mol Br}_2}{159.81 \text{ g Br}_2} \times \frac{2 \text{ mol AlBr}_3}{3 \text{ mol Br}_2} = 0.161 \text{ mol AlBr}_3$$

reveals that Br_2 is the limiting reactant. Then using this amount we can calculate the grams of $AlBr_3$.

$$0.161 \text{ mol AlBr}_3 \times \frac{266.70 \text{ g AlBr}_3}{\text{mol AlBr}_3} = 42.9 \text{ g AlBr}_3$$

Exercise 10.19

The balanced equation $3\ Ca(OH)_2 + 2\ H_3PO_4 \rightarrow Ca_3(PO_4)_2 + 6\ H_2O$ is used to compute the mol $Ca_3(PO_4)_2$ from each reactant:

$$76.5\ g\ Ca(OH)_2 \times \frac{mol\ Ca(OH)_2}{74.093\ g\ Ca(OH)_2} \times \frac{mol\ Ca_3(PO_4)_2}{3\ mol\ Ca(OH)_2} = 0.344\ mol\ Ca_3(PO_4)_2$$

$$100.0\ g\ H_3PO_4 \times \frac{mol\ H_3PO_4}{97.9951\ g\ H_3PO_4} \times \frac{mol\ Ca_3(PO_4)_2}{2\ mol\ H_3PO_4} = 0.5102\ mol\ Ca_3(PO_4)_2$$

$Ca(OH)_2$ is limiting:

$$0.344\ mol\ Ca_3(PO_4)_2 \times \frac{310.177\ g\ Ca_3(PO_4)_2}{mol\ Ca_3(PO_4)_2} = 107\ g\ Ca_3(PO_4)_2$$

Exercise 10.20

$\boxed{mL\ Br_2} \rightarrow \boxed{g\ Br_2} \rightarrow \boxed{mol\ Br_2} \rightarrow \boxed{mol\ AlBr_3} \rightarrow \boxed{g\ AlBr_3}$ yields,

$$12.8\ mL\ Br_2 \times \frac{3.12\ g\ Br_2}{mL\ Br_2} \times \frac{mol\ Br_2}{159.81\ g\ Br_2} \times \frac{2\ mol\ AlBr_3}{3\ mol\ Br_2} \times \frac{266.69\ g\ AlBr_3}{mol\ AlBr_3} = 44.4\ g\ AlBr_3$$

Exercise 10.21

First we use each reactant to calculate the mol $FeCl_3$

$$25.0\ L\ HCl \times \frac{1.47\ g\ HCl}{L} \times \frac{mol\ HCl}{36.4606\ g\ HCl} \times \frac{2\ mol\ FeCl_3}{6\ mol\ HCl} = 0.336\ mol\ FeCl_3$$

$$4.32\ g\ Fe_2O_3 \times \frac{mol\ Fe_2O_3}{159.69\ g\ Fe_2O_3} \times \frac{2\ mol\ FeCl_3}{mol\ Fe_2O_3} = 0.0541\ mol\ FeCl_3\text{, thus } Fe_2O_3 \text{ is limiting.}$$

Fe_2O_3 is limiting:

$$0.0541\ mol\ FeCl_3 \times \frac{162.21\ g\ FeCl_3}{mol\ FeCl_3} = 8.78\ g\ FeCl_3$$

Alternatively, we could have calculated the g $FeCl_3$ theoretically from each reactant and selected the smallest amount.

C. Chapter 10 Practice Test

1. Chemical equations
   a. must be balanced.
   b. have equal numbers of reactants and products.
   c. have chemical formulas as terms.
   d. represent chemical reactions.

2. Which of the following equation(s) are balanced?
   a. $Al + 3\ O_2 \longrightarrow Al_2O_3$
   b. $NH_4NO_3 \longrightarrow N_2O + 2\ H_2O$
   c. $Na_2O + H_2O \longrightarrow NaOH$
   d. $2\ FeCl_3 + 3\ Ca(OH)_2 \longrightarrow 2\ Fe(OH)_3 + 3\ CaCl_2$

3. Balance the equation for the reaction that occurs as aluminum is dissolved in sulfuric acid forming aluminum sulfate and hydrogen gas. First write formulas for reactants and products.

4. Coefficients in chemical equations can be interpreted as _______ or ______, but not_______.

5. Calculate the number of moles of hydrogen and nitrogen required to make 7.24 moles of ammonia. First write a balanced equation.

6. Write the two conversion factors relating mol H_2 and mol NH_3 in the reaction used in problem 5.

7. Acetylene, C_2H_2, is produced when calcium carbide, CaC_2, is added to water. Calcium hydroxide is also produced. After writing the balanced equation, calculate how many moles of CaC_2 are needed to completely react with 98.0 g of H_2O.

8. Calculate how many grams of acetylene are produced by adding excess water to 5.0 g of CaC_2 using the reaction in 7.

9. The combustion of ethene is described by $C_2H_4(g) + 3\ O_2(g) \rightarrow 2\ CO_2(g) + 2\ H_2O(g)$. If 2.70 mol of C_2H_4 is mixed with 6.30 mol O_2 then
   (a) identify the limiting reactant.
   (b) calculate the moles of water produced.
   (c) calculate the percentage yield if in fact 36.2 g of H_2O is produced.

10. Acetylene produced by the reaction in problem 7 has a density of 0.618 g/L. Calculate the moles of calcium carbide required to produce 72 L of C_2H_2.

D. Answers To Practice Test:

1. a, c, d

2. b, d

3. reactants: Al, H_2SO_4
   products: $Al_2(SO_4)_3$, H_2
   reaction: $2\ Al + 3\ H_2SO_4 \rightarrow Al_2(SO_4)_3 + 3\ H_2$

4. molecules (formula units), moles, mass (grams)

5. $N_2 + 3\ H_2 \rightarrow 2\ NH_3$. 3.62 mol N_2 and 10.9 mol H_2 required.

6. $\frac{3\text{ mol } H_2}{2\text{ mol } NH_3}$ and $\frac{2\text{ mol } NH_3}{3\text{ mol } H_2}$

7. $CaC_2 + 2\ H_2O \rightarrow C_2H_2 + Ca(OH)_2$. 2.72 mol CaC_2 are needed.

8. 2.03 g C_2H_2 are produced.

9. (a) O_2 limiting
   (b) 4.20 mol H_2O theoretically produced
   (c) 47.8%

10. 1.71 mol CaC_2 required.

Chapter 11

Properties and Chemical Reactions of Gases

A. Outline and Study Hints

Section 11.1

- A sample of gas is characterized by its temperature (T), pressure (p), and volume (V). Pressure is defined as force per unit area; it is measured in units of atmospheres, but other conversions are used. It is helpful to memorize the conversion factors

$$760 \text{ mm Hg} = 760 \text{ torr} = 1 \text{ atm} = 101{,}325 \text{ Pa}.$$

Section 11.2

- Kinetic molecular theory describes gas molecules in random, independent motion, separated from each other by wide distances. The molecules collide with the walls of the container and with each other.

Section 11.3

- The kinetic energy of a gas depends on the temperature only. Because they possess the same energy , heavier molecules move more slowly than lighter molecules. This leads to Graham's law of effusion.

$$\frac{V_1}{V_2} = \sqrt{\frac{m_1}{m_2}}$$

Section 11.4

- Boyle's law states that pressure and volume are inversely proportional to one another at constant temperature:

$$P_1V_1 = P_2V_2$$

- We can rearrange this equation by dividing both sides by P_1:

$$\frac{\not{P}_1V_1}{\not{P}_1} = \frac{P_2V_2}{P_1}$$

$$V_1 = \frac{P_2V_2}{P_1}$$

$$V_1 = \left(\frac{P_2}{P_1}\right) \times V_2$$

$$V_1 = V_2 \times \left(\frac{P_2}{P_1}\right)$$

We can also divide both sides of the equation by V_1:

$$P_1 = \frac{P_2V_2}{V_1} = P_2 \times \left(\frac{V_2}{V_1}\right)$$

- P_1 and P_2 must be in the same units. Likewise, V_1 and V_2 must be in the same set of units.

Section 11.5

- Avogadro's hypothesis states that equal volumes of gases contain equal moles at the same temperature and pressure conditions; the volume is directly proportional to the moles contained in a gas; $V_1 = kn_1$ and $V_2 = kn_2$ or:

$$\frac{V_1}{n_1} = \frac{V_2}{n_2} \quad \text{or} \quad V_2 = V_1 \times \frac{n_2}{n_1}$$

- These relationships need not be memorized, but you should note the "forms"

$$\text{one volume} = \left(\text{other volume}\right) \times \left(\text{ratio of moles}\right)$$

- The mole ratio is deduced by common sense: more moles means more volume .

Section 11.6

- Gay-Lussac's law states that the pressure of a gas is directly proportional to its absolute temperature :

$$P = kT \quad \text{or} \quad \frac{P_1}{T_1} = \frac{P_2}{T_2} \quad \text{or} \quad P_1 = P_2 \times \frac{T_1}{T_2}$$

- Don't forget that: °C = K + 273. We must always use the Kelvin scale when doing gas law problems.

Section 11.7

- Charles' law states that the volume of a gas is directly proportional to its absolute temperature:

$$V = kT \quad \text{or} \quad \frac{V_1}{T_1} = \frac{V_2}{T_2} \quad \text{or} \quad V_1 = V_2 \times \frac{T_1}{T_2}$$

Section 11.8

- The gas laws can be combined to yield

$$\frac{P_1V_1}{T_1} = \frac{P_2V_2}{T_2}$$

with many variants. You should "play" with this equation to rearrange it to solve for either P_1, V_1 or T_1 as demonstrated in text examples and exercises. The commutative and associative rules from algebra are helpful here.

$$\frac{P_1V_1}{T_1} = P_1 \times \frac{1}{T_1} \times V_1 \qquad \text{(commutative or order of operations)}$$

$$\frac{P_1V_1}{T_1} = P_1 \times \left(\frac{V_1}{T_1}\right) \qquad \text{(associative or grouping terms)}$$

Section 11.9

- The ideal gas law combines the gas laws:

$$PV = nRT$$

where

$$R = 0.0821 \frac{\text{L atm}}{\text{mol K}} = 62.4 \frac{\text{L torr}}{\text{mol K}}$$

depending on the units of P.

- You should be able to solve for each variable:

$$\boxed{P = \frac{nRT}{V}} \quad \boxed{n = \frac{PV}{RT}} \quad \boxed{V = \frac{nRT}{P}} \quad \boxed{T = \frac{PV}{nR}}$$

Section 11.10

- Standard temperature and pressure (STP) is 0 °C and 1 atm. The ideal gas law then tells us the standard molar volume for gases is 22.4 L/mol.

Section 11.11

- Gas density, given by mass/volume can be used with the ideal gas law to calculate

$$\text{molar mass, } M = \frac{dRT}{P}.$$

Section 11.12

- The sum of the partial pressures of all the gases in a mixture is equal to the total pressure. This is called Dalton's law of partial pressures. When gases are collected over water, the vapor pressure of water, P_{H_2O} must be subtracted from the total pressure to obtain the gas pressure.

Section 11.13

- Gas stoichiometry uses the solution map,

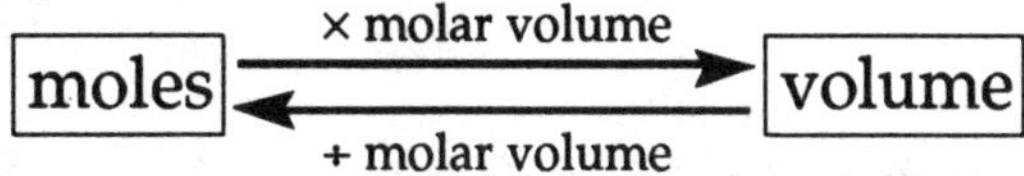

and the ideal gas law is used to determine the molar volume if STP conditions do not prevail.

- A balanced chemical equation must always be used in stoichiometry calculations. This section is a good review from the last chapter.

Section 11.14

- Gases begin to deviate from the ideal gas law at low temperatures and high pressure, due to intermolecular forces between the molecules .

B. Solutions to In-Text Exercises

Exercise 11.1

Using the conversion 760 torr = 1 atm,

$$0.455 \text{ atm} \times \frac{760 \text{ torr}}{\text{atm}} = 346 \text{ torr.}$$

Exercise 11.2

Noting that mm Hg = torr and using the conversion 760 mm Hg = 101,325 Pa,

$$642 \text{ mm Hg} \times \frac{101{,}325 \text{ Pa}}{760 \text{ mmHg}} = 8.56 \times 10^4 \text{ Pa.}$$

Exercise 11.3

Using $V_1 = V_2 \times \frac{P_2}{P_1}$ yields,

$$2.00 \text{ L} \times \frac{1.00 \text{ atm}}{0.750 \text{ atm}} = 2.67 \text{ L.}$$

Exercise 11.4

Using $P_1 = P_2 \times \frac{V_2}{V_1}$ yields,

$$900. \text{ torr} \times \frac{6.50 \text{ L}}{10.0 \text{ L}} = 585 \text{ torr.}$$

Exercise 11.5

First calculate the pressure in torr, then convert to atm:

$$695 \text{ torr} \times \frac{3.1 \text{ L}}{5.7 \text{ L}} \times \frac{\text{atm}}{760 \text{ torr}} = 0.50 \text{ atm.}$$

Exercise 11.6

Using $V_2 = V_1 \times \frac{n_2}{n_1}$ yields

$$8.74 \text{ L} \times \frac{1.25 \text{ mol}}{2.50 \text{ mol}} = 4.37 \text{ L.}$$

Exercise 11.7

Obviously, the answer must be more than 5.32 mol,

$$5.32 \text{ mol} \times \frac{30.4 \text{ L}}{18.6 \text{ L}} = 8.70 \text{ mol.}$$

Exercise 11.8

Using $T_1 = T_2 \times \frac{P_1}{P_2}$ yields

$$298\text{ K} \times \frac{5.00\text{ atm}}{1.50\text{ atm}} = 993\text{ K}.$$

Exercise 11.9

Using $P_1 = P_2 \times \frac{T_1}{T_2}$, and converting 185 °C to 458 K and 2 °C to 275 K yields

$$795\text{ torr} \times \frac{275\text{ K}}{458\text{ K}} = 477\text{ torr}.$$

Exercise 11.10

Using $V_1 = V_2 \times \frac{T_1}{T_2}$ yields,

$$5.25\text{ L} \times \frac{455\text{ K}}{298\text{ K}} = 8.02\text{ L}.$$

Exercise 11.11

Converting 0 °C to 273 K, calculating the temperature in Kelvins, then converting to Celsius yields,

$$273\text{ K} \times \frac{2.74\text{ L}}{4.28\text{ L}} = 175\text{ K};\ 175\text{ K} - 273 = -98\text{ °C}.$$

Exercise 11.12

Start with the general relationship

$$\frac{P_2V_2}{T_2} = \frac{P_1V_1}{T_1},$$

multiply both sides by T_2 and divide both sides by P_2,

$$\frac{\cancel{T_2}}{\cancel{P_2}} \times \frac{\cancel{P_2}V_2}{\cancel{T_2}} = \frac{P_1V_1}{T_1} \times \frac{T_2}{P_2}$$

then cancel terms on the left hand side and rearrange the right hand side.

$$V_2 = \frac{P_1V_1}{T_1} \times \frac{T_2}{P_2};\quad V_2 = V_1 \times \frac{P_1}{P_2} \times \frac{T_2}{T_1}$$

Exercise 11.13

We use $T_2 = T_1 \times \frac{V_2}{V_1} \times \frac{P_2}{P_1}$ after converting 21 °C to 294K.

$$21\text{ °C} + 273 = 294;\ 294\text{ K} \times \frac{600.\text{ mL}}{719\text{ mL}} \times \frac{3.50\text{ atm}}{2.15\text{ atm}} = 399\text{ K}$$

$$399\text{ K} - 273 = 126\text{ °C}$$

Exercise 11.14

We are looking for the number of moles, $n = \frac{PV}{RT}$

$$\frac{1.75 \text{ atm} \times 4.00 \text{ L}}{(0.0821 \text{ L atm/mol K}) \times 325 \text{ K}} = 0.262 \text{ mol of gas.}$$

Exercise 11.15

Solving for $T = \frac{PV}{nR}$ yields the Kelvin temperature

$$\frac{1.25 \text{ atm} \times 0.250 \text{ L}}{0.0130 \text{ mol} \times (0.0821 \text{ L atm/mol K})} = 293 \text{ K.}$$

this is then converted to Celsius.

293 K - 273 = 20 °C

Exercise 11.16

First, calculate the moles of N_2 using the molar mass;

$$55.5 \text{ g } N_2 \times \frac{\text{mol } N_2}{28.0134 \text{ g } N_2} = 1.98 \text{ mol } N_2$$

then use ideal gas law to solve for $V = \frac{nRT}{P}$

$$\frac{1.98 \text{ mol} \times (62.4 \text{ L torr/mol K}) \times 273 \text{ K}}{850. \text{ torr}} = 39.7 \text{ L.}$$

Exercise 11.17

Using the conversion factor 1 mol = 22.4 L at STP,

$$17.2 \text{ L} \times \frac{\text{mol}}{22.4 \text{ L}} = 0.768 \text{ mol.}$$

Exercise 11.18

Multiplying by the standard molar volume yields

$$5.2 \text{ mol} \times \frac{22.4 \text{ L}}{\text{mol}} = 1.2 \times 10^2 \text{ L.}$$

Exercise 11.19

First calculate moles, $n = \frac{PV}{RT}$, then divide mass by moles:

$$\frac{(0.500 \text{ atm})(1.23 \text{ L})}{(0.0821 \text{ L atm/mol K})(300. \text{ K})} = 0.0250 \text{ mol}; \frac{0.700 \text{ g}}{0.0250 \text{ mol}} = 28.0 \text{ g/mol.}$$

Exercise 11.20

$M = \frac{dRT}{P}$ thus

$$(5.28 \text{ g/L})\left[\frac{(0.0821 \text{ L atm/mol K})(273 \text{ K})}{1.00 \text{ atm}}\right] = 118 \text{ g/mol}$$

Exercise 11.21

Using $d = \frac{PM}{RT}$ with R in torr:

$$\frac{(801 \text{ torr})(44.010 \text{ g/mol})}{(62.4 \text{ L torr/ mol K})(304 \text{ K})} = 1.86 \text{ g/ L}$$

Exercise 11.22

Again using the density relationship $d = \frac{PM}{RT}$

$$\frac{(0.824 \text{ atm})(17.0304 \text{ g/mol})}{(0.0821 \text{ L atm/mol K})(298 \text{ K})} = 0.574 \text{ g/L}$$

Exercise 11.23

First, we convert mass of N_2 into moles and then atm, then we can add the P_{N_2} and P_{SO_2}.

$$15.5 \text{ g } N_2 \times \frac{\text{mol } N_2}{28.0134 \text{ g } N_2} = 0.553 \text{ mol } N_2$$

$$\frac{(0.553 \text{ mol})(0.0821 \text{ L atm/mol K})(273 \text{ K})}{7.00 \text{ L}} = 1.77 \text{ atm}$$

$1.77 \text{ atm} + 0.918 \text{ atm} = 2.69 \text{ atm} = P_{tot}$

Exercise 11.24

Converting from torr to atm,

$32 \text{ torr} \times \frac{\text{atm}}{760. \text{ torr}} = 0.042 \text{ atm}$ = vapor pressure of water.

Subtracting this from the total pressure yields P_{N_2}.

$0.750 \text{ atm} - 0.042 \text{ atm} = 0.708 \text{ atm} = P_{N_2}$

Exercise 11.25

Write the balanced chemical equation

$$CaCO_3 \rightarrow CaO + CO_2$$

Then convert $\boxed{g} \rightarrow \boxed{\text{mol } CaCO_3} \rightarrow \boxed{\text{mol } CO_2}$

$$5.20 \text{ g } CaCO_3 \times \frac{\text{mol } CaCO_3}{100.089 \text{ g } CaCO_3} \times \frac{\text{mol } CO_2}{\text{mol } CaCO_3} = 0.05195 \text{ mol } CO_2$$

And finally to volume using ideal gas law $V = \frac{nRT}{P}$

$$\frac{(0.05195 \text{ mol})(62.4 \text{ L torr/mol K})(300. \text{ K})}{730. \text{ torr}} = 1.33 \text{ L (rounded to 3 significant figures)}$$

Exercise 11.26

Writing the balanced equation

$$C_5H_{12} + 8\,O_2 \rightarrow 5\,CO_2 + 6\,H_2O$$

Using $\boxed{L} \rightarrow \boxed{mL} \rightarrow \boxed{g\ C_5H_{12}} \rightarrow \boxed{\text{mol } C_5H_{12}} \rightarrow \boxed{\text{mol } CO_2}$

$$0.265 \text{ L } C_5H_{12} \times \frac{1000\text{mL}}{\text{L}} \times \frac{0.626 \text{ g } C_5H_{12}}{\text{mL } C_5H_{12}} \times \frac{\text{mol } C_5H_{12}}{72.150\text{g } C_5H_{12}} \times \frac{5 \text{ mol } CO_2}{\text{mol } C_5H_{12}} = 11.5 \text{ mol } CO_2$$

and finally find the volume by

$$\frac{11.5 \text{ mol} \times (0.0821 \text{ L atm/mol K}) \times 298 \text{ K}}{0.950 \text{ atm}} = 296 \text{ L}$$

C. Chapter 11 Practice Test

1. If a gas is compressed from 2 L to 1 L and the temperature remains constant, then the pressure ______.

2. One standard atmosphere is approximately
   a. 100 kPa.
   b. 100 torr.
   c. 100 mm H_2O.
   d. 100 psi.

3. Gasses at the same temperature have the same _______.

4. Match the following laws with the algebraic expression(s) listed: Boyle's, Avogadro's, Gay-Lussac's, Charles'

   a. $T_1 = T_2 \times \frac{P_1}{P_2}$ d. $P_1V_1n_1 = P_2V_2n_2$

   b. $V_1 = V_2 \times \frac{T_1}{T_2}$ e. $P_1 = P_2 \times \frac{V_2}{V_1}$

   c. $V_1 = \frac{P_2V_2}{P_1}$ f. $V = kn$

5. State the combined gas law and the ideal gas law.

6. In gas calculations, the temperature must always be in _______ whereas the volume and pressure can have variable units.

7. Calculate the new volume of 2.50 L of amnestic gas originally at 760. torr after the pressure has dropped to 304 torr without any change in temperature.

8. A cylinder of compressed nitrogen has a volume of 30. L at 100. atm and 27 °C. The cylinder is cooled until the pressure is 5.0 atm. What is the new temperature of the gas in the cylinder?

9. What is the volume occupied by 0.582 mol of carbon monoxide gas at 15 °C and 622 torr? What is the gas density under these conditions?

10. A sealed 1.00 mL indestructible container filled with nitroglycerine (d = 1.59 g/mL) is detonated and the following reaction occurs:

    $C_3H_5O_9N_3$ (nitroglycerine) $\rightarrow CO_2(g) + O_2(g) + N_2(g) + H_2O(g)$

    (a) Balance the equation.
    (b) Calculate the pressure in the container at 323 °C after detonation. Note that the products are all gasses.

D. Answers To Practice Test:

1. doubles or increases be a factor of two.

2. a

3. energy or kinetic energy

4.

Boyle's	c and e
Avogadro's	f
Gay-Lussac's	a
Charles'	b

5. $\frac{P_1V_1}{T_1} = \frac{P_2V_2}{T_2}$ and PV= nRT

6. Kelvins

7. 6.25 L

8. -258 °C or 15 K

9. 16.8 L; 0.970 g/L

10. (a) $4\ C_3H_5O_9N_3\ (l) \rightarrow 12\ CO_2\ (g) + O_2\ (g) + 6\ N_2\ (g) + 10\ H_2O\ (g)$
(b) 2480 atm; 0.00700 mol of nitroglycerin produces 0.0508 moles of gas.

Chapter 12

The Condensed States of Liquids and Solids

A. Outline and Study Hints

Section 12.1

- Gases can be condensed to solids and liquids by lowering the temperature which causes a loss of kinetic energy. The fact that these condensed phases form implies that all molecules attract each other to some extent.

Section 12.2

- Atoms are held together through chemical bonds forming molecules. Molecules are held together with attractive van der Waals forces. The distinction is illustrated below for Cl_2O:

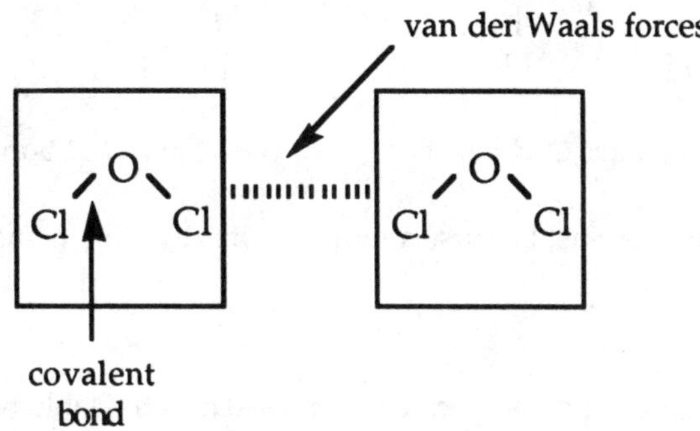

- There are two types of van der Waals forces, dipole-dipole and temporary dipole.
- Dipole-dipole forces exist between molecules that are polar. Polar molecules must have polar covalent bonds which occur when atoms of different electronegativities bond. Some examples of polar molecules are:

H−Br , $H-\ddot{N}(H)-H$, $H-C(H)(H)-\ddot{Cl}:$

- Some examples of nonpolar molecules which do not experience dipole-dipole or dipolar forces are:

$:\ddot{I}-\ddot{I}:$, $\ddot{O}=C=\ddot{O}$, CCl_4 , $H-C(H)(H)-C(H)(H)-C(H)(H)-H$

- Molecules experience temporary dipole forces, also called induced dipole forces. Molecules with more electrons experience greater dipole forces.
- A few molecules experience intermolecular forces that are stronger than van der Waals forces. These forces, called hydrogen bonds, occur in molecules that contain O-H, N-H , or F-H bonds. The H atoms on one molecule are able to get close enough to an electronegative O,N, or F on another molecule to form an attractive force:

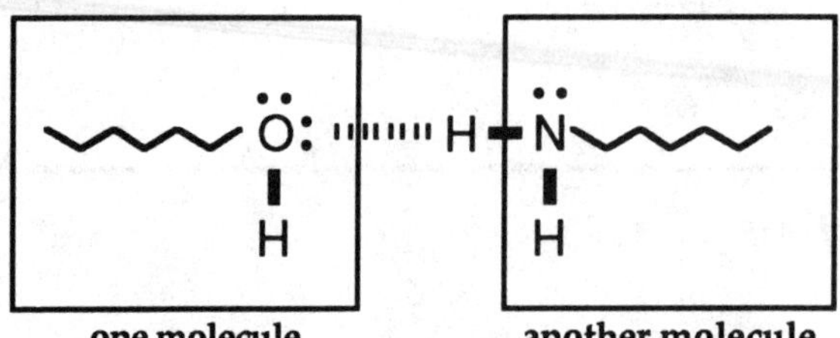

one molecule **another molecule**

- Note that you must write Lewis formulas to deduce whether or not a molecule has hydrogen bonds. For example, the two molecules with the formula C_2H_6O:

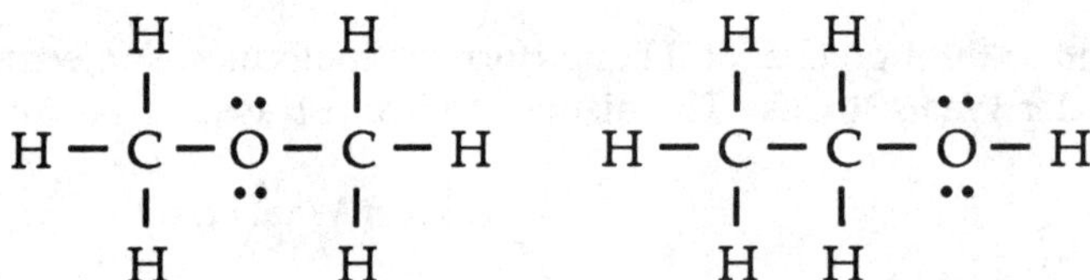

does not have O-H bond **does have O-H bond**

- The figures above illustrate that just because a molecule has hydrogen and oxygen, it may not have O-H bonds.

Section 12.3

- The properties of gases, liquids and solids can be contrasted in the table below:

property	gas	liquid	solid
density	low	high	high
flow	yes	yes	no
compressibility	high	low	low
kinetic energy	greatest	intermediate	least

Section 12.4

- Solids, liquids, and gases undergo phase changes summarized in the diagram:

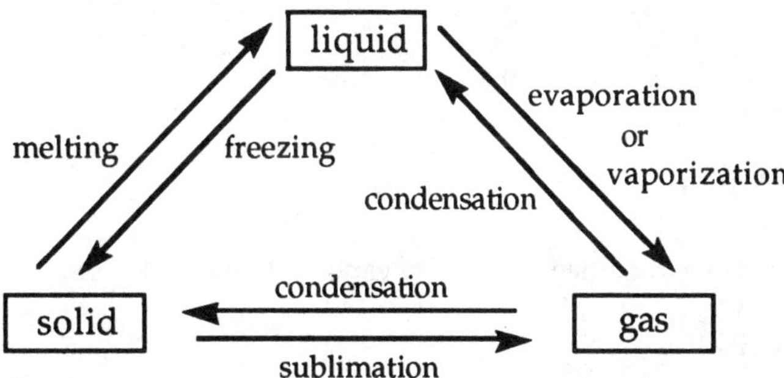

- The energy required to vaporize one gram of a substance is called the heat of vaporization. Substances with large intermolecular forces would have high heats of vaporization. Condensation is the reverse of vaporization and therefore is an exothermic process giving off heat.

Section 12.5

- When a liquid and its vapor exist together in a closed container at a constant temperature, then the rate of evaporation will eventually equal the rate of condensation and a dynamic equilibrium will be established so that the partial pressure of the vapor will be equal to the vapor pressure of the liquid. Liquid with few intermolecular forces, like butane, have large vapor pressure because few forces hold the molecules in the liquid phase.
- Vapor pressure of solids are often too small to be measured.

Section 12.6

- The temperature at which a liquid boils is called the normal boiling point. Liquids with lots of intermolecular forces have high boiling points.
- Substances that are gases at room conditions have low normal boiling points. The boiling point depends on the pressure exerted on the surface of the liquid; the lower the pressure, the lower the boiling point.

Section 12.7

- Liquids freeze as the temperature is lowered; solids melt at this same temperature. Substances with large intermolecular forces would have high melting points. The process of melting is an endothermic process; the heat required to melt one gram of a substance is called the heat of fusion.

Section 12.8

- The three categories of crystalline solids are: ionic solids, molecular solids and covalent solids. You should memorize the types of forces that hold these solids together. These are displayed in Table 12.8.

B. Solutions to In-Text Exercises

Exercise 12.1

a. Using the heat of vaporization for NH_3 from Table 12.1,

$$21.0\ \text{g} \times \frac{1.38\text{kJ}}{\text{g}} = 29.0\ \text{kJ}$$

b. First converting from kg to g, then use heat of vaporization for sodium.

$$1.35\ \text{kg} \times \frac{1000\text{g}}{\text{kg}} \times \frac{4.27\ \text{kJ}}{\text{g}} = 5760\ \text{kJ}$$

Exercise 12.2

Using the heat of fusion of sodium chloride after first converting to grams, yields

$$1.35\ \text{g} \times \frac{0.520\ \text{kJ}}{\text{g}} = 0.702\ \text{kJ}$$

C. Chapter 12 Practice Test

1. Intermolecular forces, collectively called _______ are of two types _______ and _______ plus a stronger forces called _______.

2. List the intermolecular forces that would be found in the following substances (Hint: draw Lewis diagrams): ammonia, dinitrogen monoxide, bromine, krypton, carbon dioxide.

3. Vaporization is an endothermic process that converts a _______ to a ________.

4. Calculate the energy required to vaporize a liter of ethanol, C_2H_5OH (d= 0.8 g/mL) at its boiling point.

5. The vapor pressure depends on the identity of the substance and _______ but not on _______ of substance.

6. Which of these substances is expected to have the highest melting temperature?
   a. methanol, CH_3OH
   b. methane, CH_4
   c. ethanol, CH_3CH_2OH
   d. ethane, CH_3CH_3

7. Intermolecular forces affect
   a. vapor pressure.
   b. melting point.
   c. freezing point.
   d. heat of fusion.

8. Calculate the amount of heat needed to melt three moles of sodium.

9. Identify the type of crystalline solid formed by potassium sulfate, benzene (C_6H_6), and silicon.

10. Why would strawberry farmers spray their fields with a mist of water vapor when they expect a frost?

D. Answers To Practice Test:

1. Van der Waals forces, dipole-dipole, temporary dipole-dipole, hydrogen bonding

2.

```
   ••
H — N — H
    |
    H
```

} dipole-dipole, temporary dipole-dipole, hydrogen bonding

```
••        ••
N = N = O
••        ••
```

} dipole-dipole, temporary dipole-dipole

```
  ••     ••
: Br — Br :
  ••     ••

   ••
 : Kr :
   ••

••        ••
O = C = O
••        ••
```

} temporary dipole-dipole only

3. liquid, gas or vapor.

4. 700 kJ

5. temperature, amount

6. c, largest, heaviest and has hydrogen bonding

7. all of these.

8. 7.8 kJ

9. K_2SO_4, ionic; C_6H_6, molecular (temporary dipole); Si, covalent.

10. To protect the crop from freezing; the water condenses as it is sprayed, thus giving off energy and warming the plants.

Chapter 13

Solutions

A. Outline and Study Hints

Section 13.1

- A solution consists of solvents and solutes. Aqueous solutions have water as the solvent. A solvent is the substance that has the same phase as the solution or that is present in the greatest quantity.

Section 13.2

- Most solutes do not dissolve in solvents to an infinite extent. Instead, the solubility or how much solute dissolves in a given amount of solvent, depends on the nature of the solute, the temperature and the pressure. Saturated solutions cannot dissolve any more solute. Supersaturated solutions are unstable cases where the solubility is temporarily exceeded. Solubility for solids and liquids generally increases with increasing temperature. Gas solubility decreases with increasing temperature and increases with increasing pressure.

Section 13.3

- Like dissolves like. Polar substances tend to dissolve in each other; in a similar manner, nonpolar substances dissolve in nonpolar substances. Polar and nonpolar substances usually do not dissolve in one another.

Section 13.4

- Compounds that dissociate completely or ionize completely are called strong electrolytes. Soluble salts are strong electrolytes. When the cations and anions form, they become solvated or hydrated by surrounding water molecules.
- Some acids are also strong electrolytes and dissociate or ionize completely, but most acids do not undergo ionization to a large extent and are thus called weak electrolytes. You should note that just because a solute dissolves (or is soluble) doesn't mean that it dissociates into ions (or ionizes).

Section 13.5

- Compounds that have attached water are called hydrates. The waters of hydration are denoted by a dot after the formula followed by the number of water molecules. Greek prefixes are used to state the number of water molecules. Cobalt(II) chloride pentahydrate consists of Co^{2+} and Cl^- ions and 5 H_2O; thus the formula is

$$CoCl_2 \bullet 5\,H_2O$$

- Hydrates can be heated to drive off the water leaving an anhydrous salt.

Section 13.6

- The solubility rules (Table 13.3) should be memorized. You could make flash cards with ionic formulas on the front and names and solubilities on the back with the rules and exceptions stated:

$AgC_2H_3O_2$	silver acetate → insoluble. Most acetates are soluble but Ag is an exception.
front	*back*

Some examples are:

Compound	Soluble	Reasons and Comments
$NaNO_3$	yes	Most compounds of alkali metal salts and all nitrate salts are soluble.
AgCl	no	Silver chloride is an exception to the general solubility of chlorides.
PbI_2	slightly	Because chloride and bromide lead salts are slightly soluble.
CsI	yes	Compounds of alkali metals are soluble.
$CsNO_3$	yes	Same as $NaNO_3$.
$Sr_3(PO_4)_2$	no	PO_4^{3-} is not listed; thus phosphates are insoluble.
KCl	yes	Same as CsI.
$Ba(ClO_3)_2$	yes	Chlorates are soluble.
CuS	no	Sulfides are usually insoluble.
$Mg(OH)_2$	no	Hydroxides are usually insoluble.

- These compounds are used in the examples and exercises in sections 13.6 through 13.8 in the text.
- You must be familiar with terms like alkali metals (group 1A metals) and halides (group VIIA anions) to use the rules.

Section 13.7

- Soluble compounds can react to form precipitates (insoluble salts), water (HOH), or gases (CO_2 or SO_2) by the decomposition of carbonic acid (H_2CO_3) or sulfurous acid (H_2SO_3). If no precipitate, gas, or water forms, then no reaction has taken place. For example, mixing $BaCl_2$ and $NaNO_3$

results in barium nitrate ($Ba(NO_3)_2$) and sodium chloride (NaCl), both of which are soluble salts according to the solubility rules. No reaction occurs. In equation format:

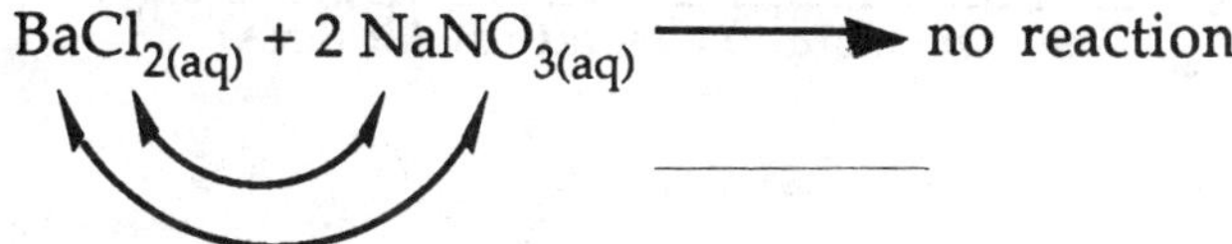

- The arrows below the reacting salts remind us that the cation on one salt combines with the anion on the other. Note that we write the formulas of products based on the charges of the combining ions; the resulting equation then must be balanced. Because the product side contains neither a precipitate, water or a gas, we say that "it doesn't go" or there is no reaction. A solid knowledge of the solubility rules is crucial to writing aqueous reactions involving salts and acids.

Section 13.8

- The net ionic equation is obtained by writing all electrolytes as ions and cancelling those ions (species) that appear on both sides of the equation.

Section 13.9

- The concentration of a solution can be expressed as a weight percent (grams of solute per 100 g of solution), molarity (moles of solute per liter of solution), or molality (moles of solute per kg of solvent). The definitions of these should be memorized; pay particular attention to the words solution and solvent.
- Concentration units can be converted using density (grams of solution per milliliter of solution). Plenty of practice is given in text exercises. The relevant equations are summarized here:

$$\text{solute wt\%} = \frac{\text{solute mass}}{\text{solution mass}} \times 100$$

$$\text{solvent wt\%} = 100 - \text{wt\% solute}$$

$$M = \frac{\text{mole solute}}{\text{L solution}} = (\text{molarity, "molar", designated [ ]})$$

this gives the solution map to inter-convert moles and volume using molarity,

$$\boxed{\text{mol}} \underset{\times\ \text{molarity}}{\overset{+\ \text{molarity}}{\rightleftarrows}} \boxed{\text{L}}$$

moles of solute — solution volume

For dilute aqueous solutions, wt% solute and M are related through density, d

$$\left(\frac{\text{wt\% solute}}{100}\right) \times (\text{density}) \times 1000 + (\text{solute molar mass}) = M$$

Molality is defined as the moles of solute per kilogram of solvent,

$$m = \frac{\text{mol solute}}{\text{kg solvent}} = (\text{molality, "molal"}).$$

wt% solute and m are related simply:

$$\frac{(\text{wt\% solute}) \times 1000}{(100 - \text{wt\% solute}) \times (\text{solute molar mass})} = m$$

Section 13.10

- Dilution occurs when a solution is mixed with solvent or another solution. Because the number of moles of solute before dilution is equal to the number of moles of solute after dilution, we have:

$$M_2 = M_1 \times \frac{V_1}{V_2}$$

where the subscript 1 refers to before dilution and 2 refers to after.

Section 13.11

- Adding a solute to a solvent lowers the vapor pressure of that solvent. This affects the colligative properties of the solution. The boiling point is elevated and the freezing point is depressed by an amount proportional to the molality:

$$\Delta T_b = m \bullet K_b \text{ and } \Delta T_f = m \bullet K_f$$

These relationships can be used to calculate the boiling and freezing points of solutions.

Section 13.12

- Solution stoichiometry uses molarity as a conversion factor. We often begin or end with a volume in these calculations. Always write a solution map before you begin a calculation.

Think moles!

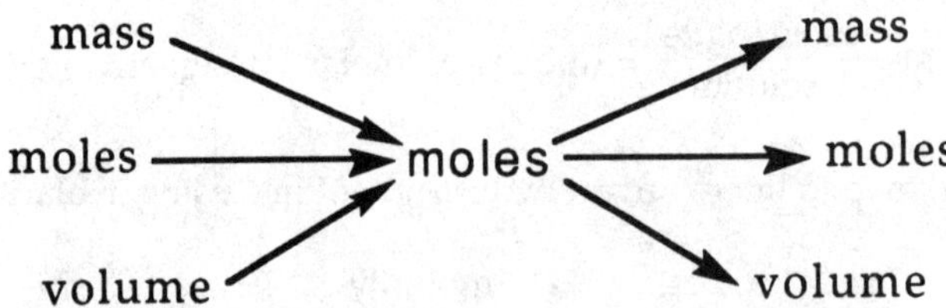

B. Solutions to In-Text Exercises

Exercise 13.1

Hg_2S is insoluble. K_2SO_4 is soluble. $CaSO_4$ is insoluble. Ag_3PO_4 is insoluble, as are most ionic compounds whose anions are not listed in Table 13.3.

Exercise 13.2

a) Reaction. $Sr_3(PO_4)_2$ is insoluble, whereas KCl(aq) is soluble.
b) Reaction. H_2O is formed, whereas $Ba(ClO_3)_2$ is soluble.
c) Reaction. CuS is insoluble, whereas $NaNO_3$(aq) is soluble.

Exercise 13.3

Writing the equation

$$2\,NaOH(aq) + MgCl_2(aq) \rightarrow Mg(OH)_2(s) + 2\,NaCl(aq)$$

and then separating into ions

$$2\,Na^+ + 2\,OH^- + Mg^{2+} + 2\,Cl^- \rightarrow Mg(OH)_2(s) + 2\,Na^+ + 2\,Cl^-$$

followed by cancelling ions yields the net ionic equation,

$$2\,OH^-(aq) + Mg^{2+}(aq) \rightarrow Mg(OH)_2(s)$$

Exercise 13.4

Total solution mass is

$$21.3\text{ g }NaH_2PO_4 + 455\text{ g }H_2O = 476\text{ g solution}$$

Dividing the solute mass by solution mass yields the weight %.

$$\frac{21.3\text{ g }NaH_2PO_4}{476\text{ g solution}} \times 100\% = 4.47\ \%\ NaH_2PO_4$$

Exercise 13.5

Solution mass times weight percent equals solute mass

$$355\text{ g} \times \frac{3.2\text{ g }C_2H_5OH}{100\text{ g solution}} = 11\text{ g }C_2H_5OH$$

Exercise 13.6

$$\boxed{\text{g solution}} \rightarrow \boxed{\text{g }SnCl_2} \rightarrow \boxed{\text{mol }SnCl_2} \rightarrow \boxed{\text{mol }SnCl_2 \bullet 2\,H_2O} \rightarrow \boxed{\text{g }SnCl_2 \bullet 2\,H_2O}$$

$$\begin{matrix}75\text{ g}\\ \text{solution}\end{matrix} \times \frac{8.0\text{ g }SnCl_2}{100\text{ g solution}} \times \frac{\text{mol }SnCl_2}{189.615\text{ g }SnCl_2} \times \frac{\text{mol }SnCl_2 \bullet 2\,H_2O}{\text{mol }SnCl_2} \times \frac{225.646\text{ g }SnCl_2 \bullet 2\,H_2O}{\text{mol }SnCl_2 \bullet 2\,H_2O} = 7.1\text{ g }SnCl_2 \bullet 2\,H_2O$$

Exercise 13.7

The wt% of solvent = 100 - wt% NaCl = 100 - 35 = 65, thus solution mass × wt% solvent yields

$$650\text{ g solution} \times \frac{65\text{ g solvent}}{100\text{ g solution}} = 420\text{ g solvent}$$

Exercise 13.8

$$\text{Molarity} = \frac{\text{moles of solute}}{\text{L of solution}}\text{; thus}$$

$$\frac{4.5\text{ mol HCl}}{35\text{ L solution}} = 0.13\text{ M HCl}$$

Exercise 13.9

First calculate mol of solute by dividing by molar mass

$$7.20\text{ g }NaC_2H_3O_2 \times \frac{\text{mol }NaC_2H_3O_2}{82.04\text{ g }NaC_2H_3O_2} = 0.0878\text{ mol }NaC_2H_3O_2$$

then calculate the liters of solution

$$500.0\text{ mL solution} \times \frac{\text{L solution}}{1000\text{ mL solution}} = 0.5000\text{ L solution}$$

finally divide to obtain solution molarity

$$\frac{0.0878\text{ mol }NaC_2H_3O_2}{0.500\text{L solution}} = 0.176\text{ M }NaC_2H_3O_2$$

Exercise 13.10

Performing the calculations on a single line,

$$\boxed{\frac{\text{g}}{\text{mL}}} \rightarrow \boxed{\frac{\text{mol}}{\text{mL}}} \rightarrow \boxed{\frac{\text{mol}}{\text{L}}}$$

yields

$$\frac{40.3\text{ g }Fe(ClO_4)_2}{750\text{ mL solution}} \times \frac{\text{mol }Fe(ClO_4)_2}{254.748\text{ g }Fe(ClO_4)_2} \times \frac{1000\text{ mL solution}}{\text{L solution}} = 0.21\text{M }Fe(ClO_4)_2.$$

Section 13.11

$$\boxed{\text{mL}} \rightarrow \boxed{\text{L}} \rightarrow \boxed{\text{mol}} \rightarrow \boxed{\text{g}}$$

$$250\text{ mL solution} \times \frac{\text{L solution}}{1000\text{ mL solution}} \times \frac{0.45\text{ mol NaOH}}{\text{L solution}} \times \frac{39.9971\text{g NaOH}}{\text{mol NaOH}} = 4.5\text{ g NaOH}$$

Exercise 13.12

$$\boxed{\text{mL solution}} \rightarrow \boxed{\text{L solution}} \rightarrow \boxed{\text{mol Cl}} \rightarrow \boxed{\text{mol }CoCl_3} \rightarrow \boxed{\text{g }CoCl_3}$$

$$137\text{ mL solution} \times \frac{\text{L solution}}{1000\text{ mL solution}} \times \frac{0.25\text{ mol }Cl^-}{\text{L solution}} \times \frac{\text{mol }CoCl_3}{3\text{ mol }Cl^-} \times \frac{165.291\text{ g }CoCl_3}{\text{mol }CoCl_3} = 1.9\text{ g }CoCl_3$$

Exercise 13.13

Each mol of Na_3PO_4 contains 3 mol Na^+, thus

$$\frac{2.0 \text{ mol } Na_3PO_4}{\text{L solution}} \times \frac{3 \text{ mol } Na^+}{\text{mol } Na_3PO_4} = 6.0 \text{ M } Na^+$$

or $[Na^+] = 6.0$ M

Exercise 13.14

$$\boxed{\frac{\text{g solute}}{\text{L solution}}} \rightarrow \boxed{\frac{\text{mol solute}}{\text{L solution}}}$$

$$\frac{119.3 \text{ g } Be_3(PO_4)_2}{1.73 \text{ L solution}} \times \frac{\text{mol } Be_3(PO_4)_2}{216.98 \text{ g } Be_3(PO_4)_2} = 0.318 \text{ M } Be_3(PO_4)_2$$

but each mol of salt has 3 mol Be^{2+}

$$\frac{0.318 \text{ mol } Be_3(PO_4)_2}{\text{L solution}} \times \frac{3 \text{ mol } Be^{2+}}{\text{mol } Be_3(PO_4)_2} = 0.954 \text{ M } Be^{2+}$$

and 2 mol PO_4^{3-}

$$\frac{0.318 \text{ mol } Be_3(PO_4)_2}{\text{L solution}} \times \frac{2 \text{ mol } PO_4^{3-}}{\text{mol } Be_3(PO_4)_2} = 0.636 \text{ M } PO_4^{3-}$$

Exercise 13.15

$$\boxed{\frac{\text{g solute}}{\text{g solution}}} \rightarrow \boxed{\frac{\text{mol solute}}{\text{g solution}}} \rightarrow \boxed{\frac{\text{mol solute}}{\text{mL solution}}} \rightarrow \boxed{\frac{\text{mol solute}}{\text{L solution}}}$$

$$\frac{35.2 \text{ g NaCl}}{100 \text{ g solution}} \times \frac{\text{mol NaCl}}{58.4425 \text{ g NaCl}} \times \frac{1.05 \text{ g solution}}{\text{mL solution}} \times \frac{1000 \text{ mL solution}}{\text{L solution}} = 6.32 \text{ M NaCl}$$

Exercise 13.16

First write the dissociation reaction

$$K_2CO_3(aq) \rightarrow 2\,K^+ + CO_3^{2-}$$

then follow the solution map for exercise 13.15

$$\frac{87.3 \text{ g } K_2CO_3}{100 \text{ g solution}} \times \frac{\text{mol } K_2CO_3}{138.2058 \text{ g } K_2CO_3} \times \frac{2 \text{ mol } K^+}{\text{mol } K_2CO_3} \times \frac{1.12 \text{ g solution}}{\text{mL solution}} \times \frac{1000 \text{ mL solution}}{\text{L solution}} = 14.1 \text{ M } K^+$$

Exercise 13.17

$$\boxed{\frac{\text{mol solute}}{\text{L solution}}} \rightarrow \boxed{\frac{\text{g solute}}{\text{L}}} \rightarrow \boxed{\frac{\text{g solute}}{\text{mL}}} \rightarrow \boxed{\frac{\text{g solute}}{\text{g solution}}} \times \boxed{100\%}$$

$$\frac{12 \text{ mol } C_2H_5OH}{\text{L solution}} \times \frac{46.069 \text{ g } C_2H_5OH}{\text{mol } C_2H_5OH} \times \frac{\text{L solution}}{1000 \text{ mL solution}} \times \frac{\text{mL solution}}{0.894 \text{ g solution}} \times 100\% = 62\% \ C_2H_5OH$$

Exercise 13.18

Convert from g Cl_2 to mol Cl_2 and g CCl_4 to kg CCl_4

$$\frac{7.9 \text{ g } Cl_2}{20.0 \text{ g } CCl_4} \times \frac{\text{mol } Cl_2}{70.9054 \text{ g } Cl_2} \times \frac{1000 \text{ g } CCl_4}{\text{kg } CCl_4} = 5.6 \text{ m } Cl_2$$ You could also do this in two steps.

Exercise 13.19

Benzoic acid is the solute and camphor is the solvent, thus

$$\frac{0.333 \text{ g benzoic acid}}{1.632 \text{ g camphor}} \times \frac{\text{mol benzoic acid}}{122.1 \text{ g benzoic acid}} \times \frac{1000 \text{ g camphor}}{\text{kg camphor}} = 1.67 \text{ m benzoic acid}$$

Exercise 13.20

The wt% of solvent is

100 g solution - 13.8 g C_6H_{14} = 86.2 g solvent

now convert to mol solute and kg solvent.

$$\frac{13.8 \text{ g } C_6H_{14}}{86.2 \text{ g solvent}} \times \frac{\text{mol } C_6H_{14}}{86.177 \text{ g } C_6H_{14}} \times \frac{1000 \text{ g solvent}}{\text{kg solvent}} = 1.86 \text{ m } C_6H_{14}$$

Exercise 13.21

For every 1000 g of solvent, there are

$$0.475 \text{ mol } C_{12}H_{22}O_{11} \times \frac{342.30 \text{ g } C_{12}H_{22}O_{11}}{\text{mol } C_{12}H_{22}O_{11}} = 163 \text{ g } C_{12}H_{22}O_{11}$$

giving a solution mass of

1000 g solvent + 163 g $C_{12}H_{22}O_{11}$ = 1163 g solution.

The wt% is

$$\frac{163 \text{ g } C_{12}H_{22}O_{11}}{1163 \text{ g solution}} \times 100\% = 14.0\%.$$

Exercise 13.22

Using the initial molarity and solution volume, the new concentration is

$$M_2 = 2.00 \text{ M} \times \frac{43 \text{ mL}}{155 \text{ mL}} = 0.55 \text{ M}.$$

Exercise 13.23

This is an example of a ten fold dilution.

$$M_2 = 0.60 \text{ M} \times \frac{50.0 \text{ mL}}{500.0 \text{ mL}} = 0.060 \text{ M}$$

Exercise 13.24

The final solution volume is

25.0 mL + 155.0 mL = 180.0 mL,

thus $M_2 = 2.50\ M \times \frac{155.0\ mL}{180.0\ mL} = 2.15\ M$.

Exercise 13.25

First calculate the mol solute in each original solution

$$34.0\ mL\ solution \times \frac{L\ solution}{1000\ mL\ solution} \times \frac{3.5\ mol\ HCl}{L\ solution} = 0.12\ mol\ HCl$$

$$78.5\ mL\ solution \times \frac{L\ solution}{1000\ mL\ solution} \times \frac{1.4\ mol\ HCl}{L\ solution} = 0.11\ mol\ HCl,$$

thus total number of moles of solute is

0.12 mol HCl + 0.11 mol HCl = 0.23 mol HCl in a solution volume of

34.0 mL solution + 78.5 mL solution = 112.5 mL solution.

$$\text{Thus } \frac{0.23\ mol\ HCl}{112.5\ mL\ solution} \times \frac{1000\ mL\ solution}{L\ solution} = 2.0\ M\ HCl.$$

Exercise 13.26

Calculate mol NaCl in the original solutions

$$4.0\ mL\ solution \times \frac{L\ solution}{1000\ mL\ solution} \times \frac{10.0\ mol\ NaCl}{L\ solution} = 0.040\ mol\ NaCl$$

$$2.0\ mL\ solution \times \frac{L\ solution}{1000\ mL\ solution} \times \frac{7.5\ mol\ NaCl}{L\ solution} = 0.015\ mol\ NaCl$$

0.040 mol NaCl + 0.015 mol NaCl = 0.055 mol NaCl

4.0 mL solution + 2.0 mL solution = 6.0 mL solution

$$\frac{0.055\ mol\ NaCl}{6.0\ mL\ solution} \times \frac{1000\ mL\ solution}{L\ solution} = 9.2\ M\ NaCl$$

Exercise 13.27

First calculate the KBr molality

$$\frac{6.2\text{ g KBr}}{42\text{ g H}_2\text{O}} \times \frac{\text{mol KBr}}{119.002\text{ g KBr}} \times \frac{1000\text{ g H}_2\text{O}}{\text{kg H}_2\text{O}} = 1.2\text{ m KBr}$$

then use $\Delta T_b = m \bullet K_b$ where $m = 2 \times$ molality

$$\Delta T_b = \frac{2\text{ mol solute ions}}{\text{mol KBr}} \times \frac{1.2\text{ mol KBr}}{\text{kg H}_2\text{O}} \times \frac{0.512\text{ deg kg H}_2\text{O}}{\text{mol solute ions}} = 1.2\text{ deg}$$

The actual boiling point is $100° + \Delta T_b$:

$100.00\ °\text{C} + 1.2\text{ deg} = 101.2\ °\text{C}$

Exercise 13.28

The boiling point elevation is

$83.5\ °\text{C} - 80.0\ °\text{C} = 3.4\text{ deg} = \Delta T_b = m \bullet K_b$; where $K_b = 2.53\text{ kg} \bullet \text{deg/mol}$

$$3.4\text{ deg} = m \times \frac{2.53\text{ deg kg benzene}}{\text{mol solute}}$$

solving for molality yields

$$m = \frac{1.3\text{ mol solute}}{\text{kg benzene}}$$

thus the mol of solute is

$$32.0\text{ g benzene} \times \frac{\text{kg benzene}}{1000\text{ g benzene}} \times \frac{1.3\text{ mol solute}}{\text{kg benzene}} = 0.042\text{ mol solute}$$

and the molar mass is the g solute divided by mol solute.

$$\frac{5.2\text{ g solute}}{0.042\text{ mol solute}} = 120\text{ g/mol}$$

Exercise 13.29

First calculate the solution molality

$$\frac{2.0\text{ kg NaCl}}{15.0\text{ kg H}_2\text{O}} \times \frac{1000\text{ g}}{\text{kg}} \times \frac{\text{mol NaCl}}{58.4525\text{ g NaCl}} = 2.3\text{ m}$$

then use $\Delta T_f = m \bullet K_f$ to calculate the freezing point depression,

$$\Delta T_f = \frac{2.3\text{ mol NaCl solute}}{\text{kg H}_2\text{O}} \times \frac{1.86\text{ deg kg H}_2\text{O}}{\text{mol solute}} \times \frac{2\text{ mol solute ions}}{\text{mol NaCl}} = 8.6\text{ deg}$$

Finally, subtract the depression point from the normal freezing point of water.

$0.00\ °\text{C} - 8.6\text{ deg} = -8.6\ °\text{C}$

Exercise 13.30

Finding ΔT_f

$6.7\ °C - 5.5\ °C = 1.2\ deg$

then using K_f for benzene; $\Delta T_f = m \bullet K_f$

$$1.2\ deg = m \times \frac{5.10\ deg\ kg\ benzene}{mol\ solute}$$

$$\text{we find } m = \frac{0.24\ mol\ solute}{kg\ benzene}$$

and the mol solute is obtained from the molality definition

$$16.8\ g\ benzene \times \frac{kg\ benzene}{1000\ g\ benzene} \times \frac{0.24\ mol\ solute}{kg\ benzene} = 0.0040\ mol\ solute$$

Dividing the solute mass by mol yields the molar mass.

$$\frac{2.3\ g\ solute}{0.0040\ mol\ solute} = 580\ g/mol$$

Exercise 13.31

The solution map:

$\boxed{g\ Cu} \rightarrow \boxed{mol\ Cu} \rightarrow \boxed{mol\ HNO_3} \rightarrow \boxed{L\ HNO_3} \rightarrow \boxed{mL\ HNO_3}$

$$2.83\ g\ Cu \times \frac{mol\ Cu}{63.546\ g\ Cu} \times \frac{8\ mol\ HNO_3}{3\ mol\ Cu} \times \frac{L\ solution}{3.50\ mol\ HNO_3} \times \frac{1000\ mL\ solution}{L\ solution} = 33.9\ mL\ solution$$

Exercise 13.32

First write a balanced equation

$3\ Zn(s) + 2\ H_3PO_4(aq) \rightarrow Zn_3(PO_4)_2(s) + 3\ H_2(g)$

then calculate mol $Zn_3(PO_4)_2$ that can be formed from each reactant

$$4.80\ g\ Zn \times \frac{mol\ Zn}{65.39\ g\ Zn} \times \frac{mol\ Zn_3(PO_4)_2}{3\ mol\ Zn} = 0.0245\ mol\ Zn_3(PO_4)_2$$

$$26.0\ mL\ solution \times \frac{L}{1000\ mL} \times \frac{5.0\ mol\ H_3PO_4}{L\ solution} \times \frac{mol\ Zn_3(PO_4)_2}{2\ mol\ H_3PO_4} = 0.065\ mol\ Zn_3(PO_4)_2$$

thus we see Zn is limiting.

Now use the smallest amount of $Zn_3(PO_4)_2$ to compute mass

$$0.0245\ mol\ Zn_3(PO_4)_2 \times \frac{386.1\ g\ Zn_3(PO_4)_2}{mol\ Zn_3(PO_4)_2} = 9.46\ g\ Zn_3(PO_4)_2$$

Exercise 13.33

Balancing the equation

$Mg(s) + 2\ HCl(aq) \rightarrow MgCl_2(aq) + H_2(g)$

Then computing mol H_2 that can be produced from Mg & HCl

$$2.50\text{ g Mg} \times \frac{\text{mol Mg}}{24.305\text{ g Mg}} \times \frac{\text{mol H}_2}{\text{mol Mg}} = 0.103\text{ mol H}_2$$

$$25.0\text{ mL solution} \times \frac{\text{L}}{1000\text{ mL}} \times \frac{12.0\text{ mol HCl}}{\text{L}} \times \frac{\text{mol H}_2}{2\text{ mol HCl}} \times = 0.150\text{ mol H}_2$$

thus Mg is limiting. Finally, using the molar volume at STP

$$0.103\text{ mol H}_2 \times \frac{22.4\text{ L H}_2}{\text{mol H}_2} = 2.31\text{ L H}_2$$

C. Chapter 13 Practice Test

1. Which of these mixtures are solutions? For those that are, list the solute(s) and solvent.
   a. air
   b. soil
   c. vodka
   d. steel

2. When a solution cannot dissolve any more solute, we say it is _____.

3. We expect that methanol, CH_3OH, would be a good solvent for _____ compounds, whereas carbon tetrachloride, CCl_4, would dissolve _______ compounds.

4. Which of the following is a strong electrolyte? For those that are, write the ions that form upon dissociation in aqueous solution. (Write formulas for all salts).
   a. barium sulfate dihydrate
   b. ammonium phosphate
   c. nickel(II) bromide
   d. mercury(I) carbonate

5. Write the reaction that occurs when aqueous solutions of 4 b and c are mixed. Also write the net ionic equation.

6. How many moles of solute are in 500. mL of 2.0 M $CaCl_2(aq)$? How many grams of $CaCl_2$ is this?

7. Calculate the mass of calcium in 5.00 L of sea water where $[Ca^{2+}] = 0.011M$.

8. Calculate the molality and freezing point of an aqueous solution of urea, CH_4N_2O, formed by dissolving 15.0 g of urea in 250 g of water.

9. How would you prepare 1.50 L of 0.250 M HNO_3 from concentrated acid, 16.0 M HNO_3?

10. How many milliliters of 2.50 M HNO_3 are required to dissolve an old copper penny that weighs 3.94 g? First balance the equation,

$Cu(s) + HNO_3(aq) \rightarrow Cu(NO_3)_2(aq) + NO(g) + 4\ H_2O(l)$

D. Answers To Practice Test:

1. a. air, solvent: N_2, solutes: O_2, CO_2 , etc.
   c. Vodka, solvent: water, solute: alcohol
   d. steel, solvent: Fe, solutes: C, Ni and/or Cr

2. saturated.

3. polar, nonpolar

4. a. $BaSO_4 \bullet 2\ H_2O$, insoluble
   b. $(NH_4)_3PO_4$, strong, $NH_4^+(aq) + PO_4^{3-}(aq)$

   c. $NiBr_2$, strong, $Ni^{2+}(aq) + 2Br^-(aq)$
   d. Hg_2CO_3, insoluble

5. $2\ (NH_4)_3PO_4(aq) + 3\ NiBr_2\ (aq) \rightarrow Ni_3(PO_4)_2(s) + 6\ NH_4Br(aq)$
   $3\ Ni^{2+}(aq) + 2\ PO_4^{3-}(aq) \rightarrow Ni_3(PO_4)_2(s)$

6. 1.0 mol $CaCl_2$; 110 g $CaCl_2$

7. 2.2 g Ca^{2+}

8. 0.94 m, freezing point = -1.8 °C

9. Dilute 23.4 mL of concentrated acid with enough water to make 1.50 L.

10. $3\ Cu(s) + 8\ HNO_3(aq) \rightarrow 3\ Cu(NO_3)_2(aq) + 2\ NO(g) + 4\ H_2O(l)$. 66.1 mL HNO_3 is required.

Chapter 14

Acids and Bases

A. Outline and Study Hints

Section 14.1

- There are three common definitions of acids and bases that are summarized in the table below:

Definition	Acid	Base	Examples
Arrhenius	H_3O^+ or H^+ producer	OH^- producer	$NaOH(aq) \rightarrow Na^+(aq) + OH^-(aq)$ $HCl(g) + H_2O(l) \rightarrow H_3O^+(aq) + Cl^-(aq)$ or $HCl(aq) \rightarrow H^+(aq) + Cl^-(aq)$
Bronsted-Lowry	proton donor	proton acceptor	$HCl(g) + NH_3(g) \rightarrow NH_4Cl(s)$ HCl: proton donor; NH_3: proton acceptor
Lewis	electron pair acceptor	electron pair donor	$H^+(aq) + NH_3(aq) \rightarrow NH_4^+(aq)$ H^+ (electron acceptor) + H–N̈–H with H below (electron donor) → $[H–N–H$ with H above and below$]^+$

- Hydronium ions, H_3O^+ are solvated hydrogen ions or protons, H^+. Water is amphoteric; it acts as both an acid and a base:

(1) $HCl(g) + H_2O(l) \rightarrow H_3O^+(aq) + Cl^-(aq)$
(H_2O: proton acceptor)

(2) $NH_3(g) + H_2O(l) \rightarrow NH_4^+(aq) + OH^-(aq)$
(H_2O: proton donor)

- In reaction (2) NH_3 is acting as a base (proton acceptor) forming the acid NH_4^+ (a proton donor).

  NH_4^+ and NH_3 are called a conjugate acid/base pair.

- Acids neutralize bases, often forming water. This can be thought of as an exchange of protons; the acid giving its proton to the base.

Section 14.2

- There are only six common strong acids; Because they ionize completely in solution, they are strong electrolytes and their solutions conduct electricity. They are listed in Table 14.1 and should be memorized (flash-cards?). You may wish to review the acid naming rules in Sections 7.3 and 7.4. Binary acids (like HCl, HF, H_2S) are named with hydro- as a prefix. The oxyacids (like HClO, H_2SO_3) have some systematic names. The vast majority of acids are "weak"; they do not completely ionize. Weak acids have an affinity for their H^+ ions; you can think of this as a love affair between an anion and a proton.
- The soluble alkali and alkaline earth hydroxides (like CsOH and $Sr(OH)_2$) are the only strong bases. Anions of weak acids (such as acetate, nitrite, carbonate) are bases because they are proton acceptors.
  A summary of all this is:

Category	Strong	Weak
acids	HCl, HBr, HI, HNO_3, $HClO_4$, H_2SO_4	most others
bases	soluble hydroxides	NH_3 and all anions except: SO_4^{2-}, ClO_4^-, NO^{3-}, I^-, Br^-, Cl^-

Section 14.3

- Ionic compounds other than acids and bases are called salts. Sometimes bases are also called salts. Neutralization reactions often produce salt and water. In the reaction between sulfuric acid and magnesium hydroxide, the H^+ attaches to the OH^- and the magnesium ion forms a salt with the sulfate:

$$H_2SO_4 + Mg(OH)_2 \rightarrow MgSO_4 + 2\ H_2O$$

Section 14.4

- Water undergoes an ionization reaction with itself represented by any of these equations:

$$H_2O \rightarrow H^+ + OH^-$$

$$\text{or } H_2O + H_2O \rightarrow H_3O^+ + OH^-$$

$$\text{or } 2\ H_2O \rightarrow H_3O^+ + OH^-$$

- This self-ionization occurs to a very small extent; the $[H_3O^+] = [\ OH^-] = 1 \times 10^{-7}$ M in pure water.

Section 14.5

- In every aqueous solution, $[OH^-]$ and $[H_3O^+]$ are inversely proportional:

$$[H_3O^+][OH^-] = 1.00 \times 10^{-14} = K_w \text{ at } 25\ °C$$

- Solutions that have $[H_3O^+] > [OH^-]$ are said to be acidic; those where $[OH^-] > [H_3O^+]$ are basic. Acids and bases added to water determine either $[OH^-]$ or $[H_3O^+]$; the other quantity is obtained by using the K_w expression. For example in a 0.0500 M HNO_3 solution, $[H_3O^+] = 0.0500$ M.

  Therefore

$$[OH^-] = \frac{K_w}{[H_3O^+]} = \frac{1.00 \times 10^{-14}}{0.0500} = 2.00 \times 10^{-13}$$

- In performing this calculation with your calculator, you would enter the numerator in exponential notation: 1, exp or EE, 14, +/-, / , .05, =.
- Your calculator should display 2.00×10^{-13} as an answer.

Section 14.6

- The negative logarithm of the $[H_3O^+]$ is the pH of the solution. The pH scale

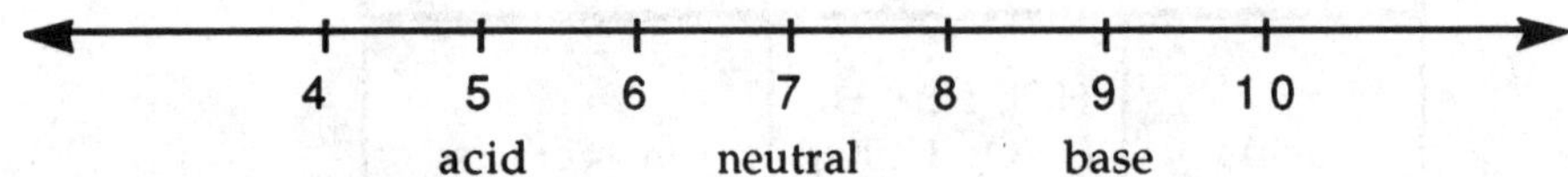

  indicates acidic, neutral and basic solutions. You will need a calculator with log and inv (or 2nd F) or 10^x buttons to do problems in this and the next section. The examples walk you through calculator use.
- A common or base 10 logarithm is the exponent to which 10 must be raised to produce a given number. For example, $x = 10^y$, then $\log x = \log(10^y) = y$. Here x is the number and y is the logarithm of x. Thus $\log 10^3 = 3$ and $-\log 10^{-2} = -(-2) = +2$.

- The reverse process of converting a logarithm into a number is called "finding the antilogarithm" or inverse log. Appropriate buttons on your calculator are "antilog", "10^x" or "inv" followed by "log". Thus antilog $0.962 = 10^{0.962} = 9.16$ and antilog $(-3.55) = 2.8 \times 10^{-4}$.

Section 14.7

- A titration involves adding one reagent solution from a buret to another solution in a flask. The volume is measured with the buret. An indicator is used to signal the end point when a slight excess of the added reagent is present. A balanced chemical equation is necessary before doing calculations.

Section 14.8

- Normality is defined as the number of equivalents per liter of solution. An equivalent refers to one mole of H_3O^+ or OH^-. One equivalent of acid reacts with on e equivalent of base.

Section 14.9

- Metal oxides and many alkali and alkaline earth metals react with water to form basic solutions. Most nonmetal oxides combine with water forming acidic solutions.

Section 14.10

- Some important industrial inorganic chemicals are H_2SO_4, NaOH, NaClO and Na_2CO_3. Sulfuric acid is manufactured from sulfur. Sodium hydroxide is obtained from NaCl(aq) and NaClO is made by reacting sodium hydroxide with chlorine gas. Soda ash, Na_2CO_3, is made from limestone.

B. Solutions to In-Text Exercises

Exercise 14.1

The proton donor is the acid; the acceptor is the base. Deprotonated acids are bases; protonated bases are acids.

$$\underset{\text{acid}}{HNO_2} + \underset{\text{base}}{OH^-} \rightarrow \underset{\text{base}}{NO_2^-} + \underset{\text{acid}}{H_2O}$$

$$\underset{\text{acid}}{H_3O^+} + \underset{\text{base}}{NH_3} \rightarrow \underset{\text{acid}}{NH_4^+} + \underset{\text{base}}{H_2O}$$

Exercise 14.2

Magnesium sulfate and water are formed; $MgSO_4$ is soluble.

$$H_2SO_4 + Mg(OH)_2 \rightarrow MgSO_4(aq) + 2\ H_2O$$

Exercise 14.3

Because $[H_3O^+] = 0.0500$ M

$$[OH^-] = \frac{1.00 \times 10^{-14}}{0.0500} = 2.00 \times 10^{-13}\ M$$

Exercise 14.4

The [NaOH] = $[OH^-]$ thus $[OH^-] = 0.00003\ M = 3 \times 10^{-5}\ M$

$$[H_3O^+] = \frac{1.00 \times 10^{-14}}{0.00003} = 3 \times 10^{-10}\ M$$

Exercise 14.5

On the calculator, entering $0.00032 = 3.2 \times 10^{-4}$, taking negative log, pH = - log 0.00032 = 3.49

Exercise 14.6

Express the pH to three decimal places

$$pH = -\log (1.59 \times 10^{-4}) = 3.799$$

Exercise 14.7

First calculate $[H_3O^+] = \frac{K_w}{[OH^-]}$; make sure that you enter 1.00×10^{-14} as 1.00, EXP or EE, 14, +/-

$$[H_3O^+] = \frac{1.00 \times 10^{-14}}{3.0 \times 10^{-3}} = 3.3 \times 10^{-12}$$

taking negative log

$pH = - \log [H_3O^+] = 11.48$

Exercise 14.8

Calculate $[H_3O^+]$ first

$$[H_3O^+] = \frac{1.00 \times 10^{-14}}{6.3 \times 10^{-5}} = 1.6 \times 10^{-10}$$

$pH = - \log (1.6 \times 10^{-10}) = 9.80$

Exercise 14.9

Enter 4.15, +/-, 10^x or antilog or inv then log:

$[H_3O^+] = 10^{-4.15} = 7.1 \times 10^{-5}$ M

Exercise 14.10

Express the concentration as a single significant figure because the pH has a single decimal place.

$[H_3O^+] = 10^{-9.2} = 6 \times 10^{-10}$ M

Exercise 14.11

First, calculate $[H_3O^+]$

$[H_3O^+] = 10^{-12.84} = 1.4 \times 10^{-13}$

then use K_w expression to obtain $[OH^-]$.

$$[OH^-] = \frac{1.00 \times 10^{-14}}{1.4 \times 10^{-13}} = 7.1 \times 10^{-2} \text{ M}$$

Exercise 14.12

Write a balanced equation (soln = solution),

$$2\ HNO_3 + Mg(OH)_2 \rightarrow 2\ H_2O + Mg(NO_3)_2$$

and calculate the mol of $Mg(OH)_2$ present .

$$18.4 \text{ mL soln} \times \frac{\text{L soln}}{1000 \text{ mL soln}} \times \frac{0.45 \text{ mol } HNO_3}{\text{L } HNO_3 \text{ soln}} \times \frac{\text{mol } Mg(OH)_2}{2 \text{ mol } HNO_3} = 4.1 \times 10^{-3} \text{ mol } Mg(OH)_2$$

Finally, use the solution volume 10.0 mL to determine the molarity.

$$\frac{4.1 \times 10^{-3} \text{ mol } Mg(OH)_2}{10.0 \text{ mL soln}} \times \frac{1000 \text{ mL soln}}{\text{L soln}} = 0.41 \text{ M } Mg(OH)_2$$

Exercise 14.13

Write a balanced equation

$$2\ HClO_4 + Ba(OH)_2 \rightarrow Ba(ClO_4)_2 + 2\ H_2O$$

then compute the mol $HClO_4$ that were present

$$500.0 \text{ mL soln} \times \frac{\text{L soln}}{1000 \text{ mL soln}} \times \frac{1.00 \text{ mol } Ba(OH)_2}{\text{L soln}} \times \frac{2 \text{ mol } HClO_4}{\text{mol } Ba(OH)_2} = 1.00 \text{ mol } HClO_4$$

which is used to compute the volume required.

$$1.00 \text{ mol } HClO_4 \times \frac{\text{L soln}}{0.500 \text{ mol } HClO_4} \times \frac{1000 \text{ mL soln}}{\text{L soln}} = 2.00 \times 10^3 \text{ mL solution}$$

Exercise 14.14

Calculate mL solution from the balanced neutralization equation.

$$HClO_3 + KOH \rightarrow KClO_3 + H_2O$$

$$12.00 \text{ mL soln} \times \frac{\text{L soln}}{1000 \text{ mL soln}} \times \frac{6.00 \text{ mol KOH}}{\text{L soln}} \times \frac{\text{mol } HClO_3}{\text{mol KOH}} = 0.0720 \text{ mol } HClO_3$$

$$0.0720 \text{ mol } HClO_3 \times \frac{\text{L soln}}{3.68 \text{ mol } HClO_3} \times \frac{1000 \text{ mL soln}}{\text{L soln}} = 19.6 \text{ mL solution}$$

Exercise 14.15

Count either the number of H^+ or OH^- that the substance provides or reacts with per formula unit. Three, one, two, one.

Exercise 14.16

Note there are 3 eq per formula unit

$$\frac{0.442 \text{ mol } H_3PO_4}{550 \text{ mL soln}} \times \frac{3 \text{ eq}}{\text{mol } H_3PO_4} \times \frac{1000 \text{ mL soln}}{\text{L soln}} = 2.41 \text{ eq/L} = 2.41 \text{ N}$$

Exercise 14.17

0.115 N = 0.115 eq/L; and there are 2 eq per mol $Mg(OH)_2$.

$$125 \text{ mL soln} \times \frac{\text{L soln}}{1000 \text{ mL soln}} \times \frac{0.115 \text{ eq}}{\text{L soln}} \times \frac{\text{mol Mg(OH)}_2}{2 \text{ eq}} = 7.19 \times 10^{-3} \text{ mol Mg (OH)}_2$$

Exercise 14.18

$\boxed{\text{g Mg(OH)}_2} \rightarrow \boxed{\text{mol Mg(OH)}_2} \rightarrow \boxed{\text{eq}}$

$$2.5 \text{ g Mg(OH)}_2 \times \frac{\text{mol Mg(OH)}_2}{58.3196 \text{ g Mg(OH)}_2} \times \frac{2 \text{ eq}}{\text{mol Mg(OH)}_2} = 0.086 \text{ eq}$$

then dividing by volume

$$\frac{0.086 \text{ eq Mg(OH)}_2}{550 \text{ mL soln}} \times \frac{1000 \text{ mL soln}}{\text{L soln}} = 0.16 \text{ N}$$

Exercise 14.19

Calculate eq

$$46.9 \text{ mL soln} \times \frac{\text{L soln}}{1000 \text{ mL soln}} \times \frac{1.66 \text{ eq}}{\text{L soln}} = 0.0779 \text{ eq}$$

then divide by volume to find normality.

$$\frac{0.0779 \text{ eq}}{25.0 \text{ mL soln}} \times \frac{1000 \text{ mL soln}}{\text{L soln}} = 3.12 \text{ N}$$

Exercise 14.20

Compute eq, then divide by volume to find N.

$$12.8 \text{ mL soln} \times \frac{\text{L soln}}{1000 \text{ mL soln}} \times \frac{3.7 \text{ eq}}{\text{L soln}} = 0.047 \text{ eq}$$

$$\frac{0.047 \text{ eq}}{10.0 \text{ mL soln}} \times \frac{1000 \text{ mL soln}}{\text{L soln}} = 4.7 \text{ N}$$

C. Chapter 14 Practice Test

1. In the reaction $HSO_4^- + H_2O \longrightarrow H_2SO_4 + OH^-$

   the proton donor is ________ whereas the proton acceptor is ________. Thus HSO_4^- is acting as a ________. The conjugate acid of HSO_4^- is ________ .

2. Identify each of the following as a strong acid, strong base, weak acid or weak base:
   a. HF
   b. $NaC_2H_3O_2$
   c. $HClO_4$
   d. NH_3

3. Write the equation for the reaction that occurs when vinegar (acetic acid) is neutralized by baking soda (sodium hydrogen carbonate) forming a salt, carbon dioxide and water.

4. Water acts as both an acid and a base, thus it is ________ . The $[OH^-]$ and $[H_3O^+]$ are________ proportional in any aqueous solution; algebraically stated _______ = K_w.

5. Calculate the $[H_3O^+]$, $[OH^-]$ and pH of a 1×10^{-3} M NaOH solution.

6. Calculate $[OH^-]$ and $[H_3O^+]$ in blood that has a pH of 7.40.

7. What is the molarity of phosphoric acid if 15.0 mL of the solution is neutralized by 38.5 mL of 0.30 M NaOH? First write a balanced equation!

8. Calculate the normality of a solution prepared by dissolving 86.3 g magnesium hydroxide in 2.5 L of solution.

9. What is the normality of an H_2SO_4 solution if 80.0 mL reacts completely with 0.424 g of Na_2CO_3? First write an equation!

10. Write the equation for the reactions that occur when
    a. Cs is dissolved in water.
    b. Sulfur dioxide is bubbled through water.
    c. Calcium oxide reacts with water.

D. Answers To Practice Test:

1. H_2O, HSO_4^-, base, H_2SO_4

2. a. weak acid
   b. weak base ($C_2H_3O_2^-$)
   c. strong acid (perchloric)
   d. weak base

3. $HC_2H_3O_2(aq) + NaHCO_3(aq) \rightarrow NaC_2H_3O_2(aq) + CO_2(g) + H_2O(l)$

4. amphoteric, inversely, $[H_3O^+]\,[OH^-] = K_w$

5. $[OH^-] = 1 \times 10^{-3}$ M, $[H_3O^+] = 1 \times 10^{-11}$ M, pH = 11.0

6. $[H_3O^+] = 4.0 \times 10^{-8}$; $[OH^-] = 2.5 \times 10^{-7}$

7. $3\,NaOH + H_3PO_4 \rightarrow Na_3PO_4 + 3\,H_2O$, 0.26 M

8. $Mg(OH)_2$, 1.2 N

9. $H_2SO_4 + Na_2CO_3 \rightarrow H_2O + CO_2 + Na_2SO_4$, 0.100 N H_2SO_4.

10. a. $Cs(s) + 2\,H_2O(l) \rightarrow CsOH(aq) + H_2(g)$
    b. $SO_2(g) + H_2O(l) \rightarrow H_2SO_3(aq)$
    c. $CaO(s) + H_2O(l) \rightarrow Ca(OH)_2(aq)$

Chapter 15

Rates of Chemical Reactions and Chemical Equilibrium

A. Outline and Study Hints

Section 15.1

- Although we have written chemical equations such as

$$H_2 + I_2 \rightarrow 2\,HI$$

the interpretation must be a conditional one: if one mole of H_2 reacts with one mole of I_2 then two moles of HI will form. This is far different than the statement: when one mole of H_2 and one mol of I_2 are mixed, then two moles of HI will form. Indeed, the stoichiometry demands the first conditional "if" statement but not the second.
- Most reactions do not proceed until one of the reactants (limiting reactant) is consumed; instead reactions are reversible which we often denote with the double arrow, $\leftrightarrows$.

Section 15.2

- Chemical reactions reach an equilibrium condition when the forward and reverse reaction rates are equal. No further charges are evident in the mixture when the system is at equilibrium.

Section 15.3

- Colliding molecules must have sufficient energy (activation energy) and the correct orientation in order for a reaction to occur where old bonds are broken and new bonds are formed.

Section 15.4

- Reaction rates are proportional to concentrations; thus removing a reactant decreases the forward reaction rate. The rate of the reverse reaction exceeds the rate of the forward reaction until equilibrium is reached again. We say the reaction has "shifted to the left" in response to the change of removal of the reactant. Addition of a reactant shifts the reaction to the right. Addition or removal of a product produces a shift to the left and right, respectively.

Section 15.5

- Increasing the temperature of a reaction mixture increases the number of collisions as well as the fraction of the molecules with sufficient activation energy, thus the rate of both the forward and reverse reactions increase. The effect on the equilibrium depends on whether the reaction is endothermic or exothermic:

Effect of	endothermic, $\Delta H > 0$	exothermic, $\Delta H < 0$
increasing temperature	shifts to right $\rightarrow$	shifts to left $\leftarrow$
decreasing temperature	shifts to left $\leftarrow$	shifts to right $\rightarrow$

Section 15.6

- Recall that the partial pressure of a gas is directly proportional to its concentration, P= [ ]RT. The concentration of a gas can thus be increased by an increase in pressure of gas or a decrease in container volume. A reaction favors the side with the fewest moles of gases if the volume is decreased, thus increasing the total system pressure.

Section 15.7

- The principles from the previous three sections are summarized in Le Chatelier's Principle: an equilibrium will shift in a direction to counteract a change and attain a new equilibrium.

Section 15.8

- Catalysts, also called enzymes in biological systems, speed up the rates of forward and reverse reactions, but they have no effect on the equilibrium.

Section 15.9

- Reactions reach equilibrium so that the equilibrium equation is obeyed. This relationship equates the concentrations of products divided by reactants to the equilibrium constant (K) which is a unique number for each reaction at a given temperature. In performing calculations, you should remember that the order of multiplication and division doesn't matter.
  For example,

$$\frac{(0.996)^2}{(9.50)\,(0.00194)} = \frac{(0.992)}{(0.01843)} = 53.8$$

or

$$= 0.996 \times 0.996 \div 9.50 \div 0.00194 = 53.8$$

or also
(you should verify)

$$= \left(0.996 \div 9.50\right) \times \left(0.996 \div 0.00194\right) = 53.8$$

Section 15.10

- Huge equilibrium constants mean that the forward reaction proceeds to a great extent before equilibrium is attained. We say that such reactions "go to completion"; very little limiting reactant(s) remains. Reactions with tiny equilibrium constants proceed to very small extents. We often say that such reactions "don't go".

Section 15.11

- When equilibrium constants are small, so little of the reactants are consumed that it is a good approximation that the reactant concentrations at equilibrium are equal to their initial concentrations.

Section 15.12

- Ionization constants for weak acids and bases are usually small equilibrium constants, K_a and K_b. An example is the general weak acid, HA ionization reaction with water:

$$HA + H_2O \leftrightarrows H_3O^+ + A^-$$

$$\frac{[H_3O^+][A^-]}{[HA]} = K_a$$

Section 15.13

- Calculations involving acids and bases can be done with the aid of a diagram shown below for Example 15.10 in the text:

$$HOCl + H_2O \leftrightarrows H_3O^+ + OCl^-$$

[initial]	$\frac{0.030}{3}$	--		0		0
[change]	- x	--		+ x		+ x
[equilibrium]	0.010 - x	--		x		x

- The [change] is the mol/L of HOCl that ionize which must be equal to the mol/L of H_3O^+ that form. We have chosen the convenient variable x to represent this concentration. The [equilibrium] entries are obtained by summing the [initial] and [change] for each species.

- Now use the [equilibrium] values in the equilibrium equation:

$$K_a = 3.5 \times 10^{-8} = \frac{[H_3O^+][OCl^-]}{[HOCl]} = \frac{x \bullet x}{0.010 - x}$$

which is now solved for x. But first note that because K_a in small we expect x to be much smaller than 0.010 so we can neglect x in the denominator. This results in a much simpler equation.

$$3.5 \times 10^{-8} \approx \frac{x^2}{0.010}$$

Section 15.14

- A buffer is a solution that has significant (at least 0.010 M) concentrations of weak acid and weak base. Addition of an acid or a base to this solution results in only a small change in pH. Addition of an acid (H_3O^+) to a buffer protonates the weak base producing more weak acid. Addition of a base (OH^-) deprotonates the weak acid forming more weak base.

Section 15.15

- Slightly soluble and "insoluble" salts dissociate to a very small extent. The solubility product is a rough measure of this dissociation. Some salts, such as $MgCO_3$, $SrCrO_4$, CuBr, $PbCO_3$, dissociate to form equal numbers of metal ions and anions, thus

$$[\text{metal ion}] = [\text{anion}]$$

and the equilibrium equation

$$[\text{metal ion}][\text{anion}] = K_{sp}$$

becomes

$$[\text{metal ion}]^2 = x^2 = K_{sp}.$$

- The above equations are not appropriate for insoluble salts such as Li_2CO_3, Ag_2SO_4, and CaF_2. Note that these salts dissociates into unequal amounts of metal ions and anions. Take Li_2CO_3 as an example:

the reaction is $\quad Li_2CO_3\,(s) \leftrightarrows 2\,Li^+ + CO_3^{2-}$

and $\quad [Li^+] = 2[CO_3^{2-}]$ or $\frac{[Li^+]}{2} = [CO_3^{2-}]$

the equilibrium equation is $\quad [Li^+]^2\,[CO_3^{2-}] = K_{sp}$

thus $\quad [Li^+]^2\,\frac{[Li^+]}{2} = \frac{[Li^+]^3}{2} = K_{sp}.$

These equations hold for Ag_2SO_4 but notice that CaF_2 has a different metal ion to anion ratio than the other two.

B. Solutions to In-Text Exercises

Exercise 15.1

The concentrations of CO_2 and H_2 will increase while the concentration of CO will decrease as the reaction shifts to the right to use up some of the added water. Not enough reaction will occur to use up all of the added H_2O, therefore, the new concentration of $H_2O(g)$ will be more than the original.

Exercise 15.2

a. reactant added, right
b. reactant removed, left
c. product removed, right
d. reactant and product added, neither (the two effects cancel each other)

Exercise 15.3

The concentration of HBr will decrease. The equilibrium will shift to the right. Increasing T causes the reaction to absorb more energy.

Exercise 15.4

The equilibrium will shift to the left, because the left has only one mole of gas whereas the right has two moles of gases. More moles of gas produces more pressure.

Exercise 15.5

The number of moles of gas is equal on both sides of the equation. The equilibrium will not shift in either direction.

Exercise 15.6

Dividing the product concentrations by reactant concentrations

$$K = \frac{[NOCl]^2}{[NO]^2[Cl_2]}\,.$$

Exercise 15.7

$$K = \frac{[O_3]^2}{[O_2]^3}$$

Exercise 15.8

The concentration is the mol divided by volume in liters.

$$[O_2] = \frac{1.00\text{ mol}}{2.00\text{ L}} = 0.500\text{ M}$$

This is substituted into the equilibrium equation

$$K = \frac{[O_3]^2}{[O_2]^3} = \frac{[O_3]^2}{(0.500)^3} = 1 \times 10^{-48}\,.$$

Multiplying both sides by $(0.500)^3 = 0.5 \times 0.5 \times 0.5 = 0.125$ or press 0.5, y^x, 3, and = on your calculator.

$$[O_3]^2 = (1 \times 10^{-48})(0.500)^3 = 1.25 \times 10^{-49}$$

$[O_3] = \sqrt{1.25 \times 10^{-49}} = 3.5 \times 10^{-25}$ which can be rounded to 3×10^{-25}

Exercise 15.9

The $[H_3O^+]$ is the antilog of -pH, thus

$$[H_3O^+] = 10^{-7.68} = 2.1 \times 10^{-8}$$

which is used in the equilibrium equation $K_a = \dfrac{[PO_4^{3-}][H_3O^+]}{[HPO_4^{2-}]}$

$$K_a = \frac{(2.1 \times 10^{-8})(2.1 \times 10^{-8})}{0.0010} = 4.4 \times 10^{-13}\text{ M}$$

Exercise 15.10

$K_w = [H_3O^+][OH^-] = (1.00 \times 10^{-7})(1.00 \times 10^{-7}) = 1.00 \times 10^{-14}$

Exercise 15.11

First write the balanced equation

$$HC_2H_3O_2 + H_2O \leftrightarrows H_3O^+ + C_2H_3O_2^-$$

[initial]	0.55	--	0	0
[change]	-x	--	+x	+x
[equilibrium]	0.55 - x	--	x	x

$$K_a = \frac{[H_3O^+][C_2H_3O_2^-]}{[HC_2H_3O_2]}$$

then inserting values from the table

$$K_a = 1.8 \times 10^{-5} = \frac{x^2}{0.55 - x}$$

Ignore x in the denominator

$x^2 = (1.8 \times 10^{-5})(0.55) = 9.9 \times 10^{-6}$

and take the square root of both sides

$x = [H_3O^+] = \sqrt{9.9 \times 10^{-6}} = 3.1 \times 10^{-3}$ M

Exercise 15.12

The reaction is the ionization of ammonia

(a) $NH_3 + H_2O \leftrightarrows NH_4^+ + OH^-$

(b) 0.08 moles of base react with the weak acid NH_4^+ to give H_2O and NH_3; the pH shifts slightly to a higher value.

(c) 0.05 moles of H_3O^+ react with the weak base NH_3 to give NH_4^+; the pH shifts to a slightly lower value.

Exercise 15.13

The dissociation equation

$PbCrO_4 \leftrightarrows Pb^{2+} + CrO_4^{2-}$

requires that

$[CrO_4^{2-}] = [Pb^{2+}] = 1.4 \times 10^{-7}$ M.

The solubility product is thus

$[CrO_4^{2-}][Pb^{2+}] = K_{sp} = (1.4 \times 10^{-7})(1.4 \times 10^{-7}) = 2.0 \times 10^{-14}$.

Exercise 15.14

The reaction is

$CuI \leftrightarrows Cu^{+} + I^{-}$ $\quad K_{sp} = [Cu^{+}][I^{-}]$

and the $[Cu^{+}] = [I^{-}] = x$, thus

$K_{sp} = 4 \times 10^{-5} = x^2$; $[Cu^{+}] = x = \sqrt{4 \times 10^{-5}} = 6 \times 10^{-3}$ M

C. Chapter 15 Practice Test

1. When a reaction has reached equilibrium
   a. ΔH = activation energy.
   b. forward reaction rate = reverse reaction rate.
   c. the equilibrium equation is obeyed.
   d. [reactants] = [products].

2. Some reactions have huge equilibrium constants and these reactions ________ .

3. Consider the decomposition of hydrogen peroxide: $2\ H_2O_2 \leftrightarrows 2\ H_2O + O_2$
   The correct interpretation of this reversible reaction is
   a. When two moles of H_2O_2 decompose, two moles of water are formed and one mol of O_2 is formed.
   b. Two moles of H_2O_2 will decompose instantaneously to two moles of H_2O and one mole of O_2.

   c. At equilibrium $[H_2O] = [O_2]$.

   d. At equilibrium $\frac{[H_2O]^2[O_2]}{[H_2O_2]^2} = K$.

4. Catalysts increase ______ but have no effect on ______ .

5. Reaction rates depend on temperature, ______ and ______ .

6. Analysis of a mixture of nitrogen, hydrogen, and ammonia in a 1.0-L flask at 400 °C shows that $[N_2] = 0.25$ M, $[H_2] = 0.15$ M and $[NH_3] = 0.10$ M.
   a. Write the balanced equation for the reaction that forms NH_3 from N_2 and H_2.
   b. Write the equilibrium equation for this reaction.
   c. Calculate the equilibrium constant for this formation reaction.

7. How is the endothermic reaction

   $C(s) + H_2O(g) + \text{heat} \leftrightarrows CO(g) + H_2(g)$

   affected by
   a. lowering the temperature
   b. decreasing the volume
   c. removing hydrogen gas from the mixture.

8. Write the chemical equation for the reaction that occurs when NaOH is added to a NaH_2PO_4/Na_2HPO_4 buffer.

9. Calculate the concentration of lead(II) in a solution saturated with PbS ($K_{sp} = 3 \times 10^{-28}$).

10. The pH of 0.50 M HNO_2 is 1.83. Calculate the ionization constant for nitrous acid.

D. Answers To Practice Test:

1. b and c

2. go to completion or occur to a large extent.

3. a and d

4. reaction rate, equilibrium or equilibrium constant.

5. concentrations, gas partial pressures

6. a. $3\ H_2(g) + N_2(g) \leftrightarrows 2\ NH_3(g)$

   b. $K = \dfrac{[NH_3]^2}{[H_2]^3[N_2]}$

   c. $K = 12$

7. a. $\leftarrow$, equilibrium shifts to the left-reactants are favored
   b. $\leftarrow$, equilibrium shifts to the left-reactants are favored
   c. $\rightarrow$, equilibrium shifts to the right-products are favored

8. $OH^- + H_2PO_4^- \rightarrow H_2O + HPO_4^{2-}$

9. $PbS \leftrightarrows Pb^{2+} + S^{2-}$, $[Pb^{2+}] = 2 \times 10^{-14}$ M

10. $HNO_2 + H_2O \leftrightarrows H_3O^+ + NO_2^-$,

    note that $[NO_2^-] = [H_3O^+]$

$$K_a = \frac{[H_3O^+][NO_2^-]}{[HNO_2]} = \frac{(10^{-pH})^2}{0.50} = 4.4 \times 10^{-4}$$

Chapter 16

Oxidation-Reduction Reactions

A. Outline and Study Hints

Section 16.1

- Electron transfer reactions, also called oxidation-reduction (redox) reactions involve electrons gained by one atom (reduction) and electrons lost by another atom (oxidation). Reduction results in a decrease of oxidation number ("charge"), whereas oxidation results in an increase in oxidation number. You should review Section 7.1 if you've forgotten how to assign oxidation numbers. Some examples are shown below:

Compound	Oxidation states of elements
Na_2SO_3	Na^+, SO_3^{2-} (S, +4; O, -2)
PCl_5	Cl, -1; P, +5
$Co(NO_2)_2$	Co^{2+}, NO_2^- (N, +3; O, -2)

- Species or compounds that are oxidized are called reducing agents; species or compounds that are reduced are called oxidizing agents. Say this statement over and over as in a mantra.
- These ideas are summarized in the reaction of magnesium with sulfur:

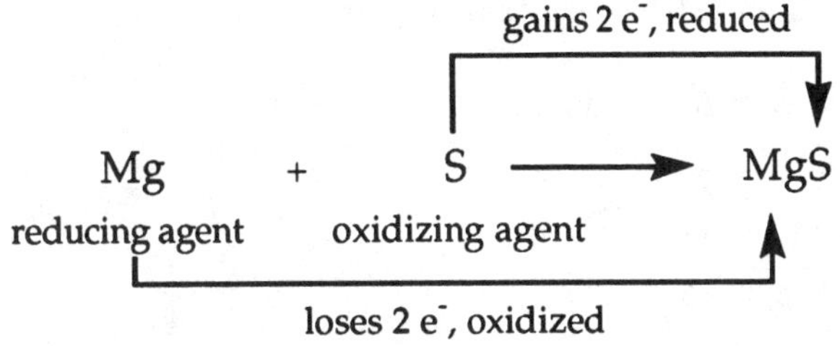

Section 16.2

- The trial and error method of balancing equations is too time-consuming for many redox reactions. There are two common methods that are useful-using oxidation numbers and using acid or base. The detailed steps are outlined in the text. Both methods require that you separate the reaction into half-reactions–one for oxidation and one for reduction. Each method also involves a step where electrons are used to "balance the charge on both sides of the equation".

For example, the equation (with oxidation states shown below the atoms)

$$\underset{+1\ -2}{H_3O^+} + \underset{+5\ -2}{IO_3^-} + \underset{+2}{Sn^{2+}} \rightarrow \underset{+4}{Sn^{4+}} + \underset{-1}{I^-} + \underset{+1\ -2}{H_2O}$$

involves the reduction half-reactions

(1) oxidation number method

$I^{5+} \rightarrow I^-$

(2) acid or base method

$IO_3^- \rightarrow I^-$

- In the oxidation number method, six electrons are added to the left side so that the charge on both sides become equal (to -1):

  $I^{5+} + 6\,e^- \rightarrow I^-$

  $+5 + (-6) = -1$

- In the acid method, we first balance the H and O using hydronium and water respectively:

  $6\,H^+ + IO_3^- \rightarrow I^- + 3\,H_2O$

  $6(+1) + (-1) \quad -1$

- The charge on the left side is +5 whereas the charge on the right is -1. We thus need to add $6e^-$ to the left to make both sides balanced (= -1 in this case):

  $6\,H^+ + IO_3^- + 6\,e^- \rightarrow I^- + 3\,H_2O$

- Only plenty of monotonous practice will result in proficiency with these methods. Religiously adhere to the order in applying the 6 and 8 steps of the two methods. Turn your paper sideways and use a full sheet for each problem. You'll save valuable time if you make no small mistakes your first time through!

Section 16.3

- Some important redox reactions that are important enough to recognize include:

  (1) use of active metals (Na, Li, Al, Zn) as reducing agents producing cations

  (2) use of dichromate ($Cr_2O_7^{2-} \rightarrow Cr^{3+}$), permanganate ($MnO_4^- \rightarrow Mn^{2+}$) and nitrate

  ($NO_3^- \rightarrow NO_2$ or NO or NH_4^+) as oxidizing agents.

  (3) combustion reactions (fuel + oxidizer)

  (4) Haber ammonia synthesis

  (5) Ostwald nitric acid synthesis

  (6) water gas reaction to make H_2

  (7) fermentation of sugars

  (8) respiration and photosynthesis

  (9) nitrogen fixation

  (10) roasting of ores and reduction with coke

  (11) corrosion; oxidation by atmosphere

Section 16.4

- If a spontaneous redox reaction can be physically separated into half reactions, then electrons will flow from the anode through an external circuit to the cathode. Such a device is called a battery.

Section 16.5

- An electrolytic cell is a battery that is forced to run in the reverse direction by applying energy in the form of electric current.
- Electrolysis is thus used to generate useful chemicals, and electroplating is used to deposit metals on surfaces.

B. Solutions to In-Text Exercises

Exercise 16.1

a. (1) Assign oxidation states to all the elements:

$Cu^{2+} + CN^- \rightarrow Cu^{1+} + (CN)_2$

+2   +2 -3   +1   +2 -2

Note that C is assigned a +2 state and N is deduced from this assignment.

(2) Write the half reactions, including only species that change oxidation numbers; balance elements.

$Cu^{2+} \rightarrow Cu^{1+}$ (reduction)

$2\ CN^- \rightarrow (CN)_2$ (oxidation)

(3) Add electrons to make the charges balance.

$Cu^{2+} + e^- \rightarrow Cu^{1+}$ (+1 on each side)

$2\ CN^- \rightarrow (CN)_2 + 2\ e^-$ (-2 on each side)

(4) Multiply reduction half reaction by 2 to consume 2 e^- that are produced in the oxidation.

$2\ (Cu^{2+} + e^- \rightarrow Cu^+)$

or

$2\ Cu^{2+} + 2\ e^- \rightarrow 2\ Cu^{1+}$

$2\ CN^- \rightarrow (CN)_2 + 2\ e^-$

(5) Add the two half reaction together and cancel like terms.

$2\ Cu^{2+} + \cancel{2e^-} + 2\ CN^- \rightarrow 2\ Cu^{1+} + (CN)_2 + \cancel{2e^-}$

(6) Check that equal numbers of atoms are on both sides and the charge is equal on both sides (+2 in this case).

$2\ Cu^{2+} + 2\ CN^- \rightarrow 2\ Cu^{1+} + (CN)_2$

b. (1) $P + Br_2 \rightarrow PBr_3$

0   0   +3 -1

(2) $P \rightarrow P^{3+}$ (oxidation)

$Br_2 \rightarrow 2\ Br^-$ (reduction)

(3) $P \rightarrow P^{3+} + 3\ e^-$

$Br_2 + 2\ e^- \rightarrow 2\ Br^-$

(4) $2\ (P \rightarrow P^{3+} + 3\ e^-)$

$3\ (Br_2 + 2\ e^- \rightarrow 2\ Br^-)$

(5) $2\ P + 3\ Br_2 + \cancel{6e^-} \rightarrow 2\ P^{3+} + 6\ Br^- + \cancel{6e^-}$

(6) $2\ P + 3\ Br_2 \rightarrow 2\ PBr_3$

Exercise 16.2

Applying steps 1 through 8 in each case:

a.

Step	Oxidation	Reduction
(1)	$Sn^{2+} \rightarrow Sn^{4+}$	$ClO_4^- \rightarrow Cl^-$
(2)	$Sn^{2+} \rightarrow Sn^{4+}$	$ClO_4^- \rightarrow Cl^-$
(3)	$Sn^{2+} \rightarrow Sn^{4+}$	$ClO_4^- + H^+ \rightarrow Cl^- + H_2O$
(4)	$Sn^{2+} \rightarrow Sn^{4+}$	$ClO_4^- + 8\ H^+ \rightarrow Cl^- + 4\ H_2O$
(5)	$Sn^{2+} \rightarrow Sn^{4+} + 2\ e^-$	$ClO_4^- + 8\ H^+ + 8\ e^- \rightarrow Cl^- + 4\ H_2O$
(6)	$4\ Sn^{2+} \rightarrow 4\ Sn^{4+} + 8\ e^-$	$ClO_4^- + 8\ H^+ + 8\ e^- \rightarrow Cl^- + 4\ H_2O$
(7)	$4\ Sn^{2+} + 8\ H^+ + ClO_4^- \rightarrow 4\ Sn^{4+} + Cl^- + 4\ H_2O$	
(8)	Check atoms and charge	

b.

Step	Oxidation	Reduction
(1)	$Cu \rightarrow Cu^{2+}$	$H_2AsO_4^- \rightarrow AsH_3$
(2)	Both reactions are balanced with regard to Cu and As.	
(3)	$Cu \rightarrow Cu^{2+}$	$H_2AsO_4^- + H^+ \rightarrow AsH_3 + H_2O$
(4)	$Cu \rightarrow Cu^{2+}$	$H_2AsO_4^- + 9\ H^+ \rightarrow AsH_3 + 4\ H_2O$
(5)	$Cu \rightarrow Cu^{2+} + 2\ e^-$	$H_2AsO_4^- + 9\ H^+ + 8\ e^- \rightarrow AsH_3 + 4\ H_2O$
(6)	$4\ Cu \rightarrow 4\ Cu^{2+} + 8\ e^-$	
(7)	$4\ Cu + H_2AsO_4^- + 9\ H^+ \rightarrow AsH_3 + 4\ Cu^{2+} + 4\ H_2O$	
(8)	Could add 9 H_2O to both sides to have H_3O^+ rather than H^+	

Exercise 16.3

In this case, we use OH^- and H_2O rather than H^+ and H_2O because the solution is basic

Step	Oxidation	Reduction
(1)	$NH_3 \rightarrow N_2$	$CuO \rightarrow Cu$
(2)	$2\ NH_3 \rightarrow N_2$	already balanced
(3) & (4)	$2\ NH_3 + 6\ OH^- \rightarrow N_2 + 6\ H_2O$	$CuO + H_2O \rightarrow Cu + 2\ OH^-$
(5)	$2\ NH_3 + 6\ OH^- \rightarrow N_2 + 6\ H_2O + 6\ e^-$	$CuO + H_2O + 2\ e^- \rightarrow Cu + 2\ OH^-$
(6)	$2\ NH_3 + 6\ OH^- \rightarrow N_2 + 6\ H_2O + 6\ e^-$	$3\ CuO + 3\ H_2O + 6\ e^- \rightarrow 3\ Cu + 6\ OH^-$
(7)	$2\ NH_3 + 6\ \cancel{OH^-} + 3\ CuO + 3\ H_2O + \cancel{6e^-} \rightarrow 3\ Cu + 6\ H_2O + N_2 + 6\ \cancel{OH^-} + \cancel{6e^-}$	
(8)	3 Waters cancel on each side, yielding $2\ NH_3 + 3\ CuO \rightarrow 3\ Cu + N_2 + 3\ H_2O$	

Exercise 16.4

Now that we are experts, we balance each half reaction applying steps (1) through (5)

$2\ (3\ OH^- + NaOH + Al \rightarrow NaAl(OH)_4 + 3\ e^-)$ (oxidation)

$3\ (2\ e^- + 2\ H_2O \rightarrow H_2 + 2\ OH^-)$ (reduction)

Now, multiply the reduction by 3 and the oxidation by 2 and add the half reactions, canceling 6 OH^- and 6 e^- appearing on both sides

$2\ NaOH + 2\ Al + 6\ H_2O \rightarrow 2\ NaAl(OH)_4 + 3\ H_2$.

If you got this one correct without help, you are doing great. Reward yourself!

C. Chapter 16 Practice Test

1. Assign oxidation numbers to all the atoms in

   $Zn + MnO_2 + NH_4Cl \rightarrow ZnCl_2 + Mn_2O_3 + NH_3 + H_2O$

2. What element(s) in the above reaction is(are)
   a. losing electrons?
   b. undergoing an increase in oxidation number?
   c. gaining electrons?
   d. oxidized?

3. In a redox reaction, the oxidizing agent _____, whereas the reducing agent _____ .

4. Identify the oxidizing agent and reducing agent in 1 above.

5. Which of the following are not electron transfer reactions?

   a. $H_2O_2 \rightarrow H_2O + O_2$

   b. $Ag_2SO_4 \rightarrow 2\ Ag^+ + SO_4^{2-}$

   c. $Cr_2O_7^{2-} + H_2O \rightarrow 2\ CrO_4^{2-} + 2H^+$

   d. $C + O_2 \rightarrow CO_2$

6. Identify the element oxidized, the element reduced, the oxidizing agent and the reducing agent for the reaction occurring when hydrosulfuric acid reacts with nitric acid producing sulfur, nitrogen monoxide gas and water. Write a balanced equation for this reaction.

7. Balance the reaction $PH_3 + I_2 + H_2O \rightarrow H_3PO_2 + HI$.

8. Write balanced half reactions for the oxidation of sulfur with nitric acid producing sulfur dioxide, nitrogen monoxide and water. Also write the balanced equation for the entire reaction.

9. Balance the reaction shown below that occurs in basic solution:

$Cr^{3+} + ClO^{-} \rightarrow CrO_4^{2-} + Cl^{-}$

10. Electrolysis of water produces _____ at the anode and hydrogen at the _______ .

D. Answers To Practice Test:

1. $\underset{0}{Zn} + \underset{+4\ -2}{MnO_2} + \underset{-3\ +1\ -1}{NH_4Cl} \rightarrow \underset{+2\ -1}{ZnCl_2} + \underset{+3\ -2}{Mn_2O_3} + \underset{-3\ +1}{NH_3} + \underset{+1\ -2}{H_2O}$

2. a. Zn b. Zn c. Mn d. Zn

3. is reduced or gains electrons, is oxidized or loses electrons

4. MnO_2 is the oxidizing agent; Zn is the reducing agent.

5. b and c

6. $3\ H_2S + 2\ HNO_3 \rightarrow 3\ S + 2\ NO + 4\ H_2O$
   S: oxidized — HNO_3: oxidizing agent
   N: reduced — H_2S: reducing agent

7. $PH_3 + 2\ I_2 + 2\ H_2O \rightarrow H_3PO_2 + 4\ HI$

8. $2\ H_2O + S \rightarrow SO_2 + 4\ H^+ + 4\ e^-$ (oxidation)
   $3\ H^+ + HNO_3 + 3\ e^- \rightarrow NO + 2\ H_2O$ (reduction)
   $4\ HNO_3 + 3\ S \rightarrow 3\ SO_2 + 4\ NO + 2\ H_2O$ (overall)

9. $2\ Cr^{3+} + 10\ OH^- + 3\ ClO^- \rightarrow 2\ CrO_4^{2-} + 5\ H_2O + 3\ Cl^-$

10. oxygen, cathode

Chapter 17

Carbon and the Compounds of Carbon

A. Outline and Study Hints

Section 17.1

- The most important property of carbon is its intermediate electron affinity and ionization energy.
- These properties allow carbon to form strong covalent bonds with a wide assortment of atoms.

Section 17.2

- **Hydrocarbons** are organic compounds that contain only hydrogen and carbon atoms.
- An hydrocarbon that does not contain any triple or double bonds is called an **alkane** or **saturated.** Remember when carbon utilizes all its four valence electrons to bond with hydrogen, it becomes saturated with hydrogen atoms.
- Methane,CH_4, is the simplest of the alkanes. The single carbon is saturated with four hydrogen atoms.
- Each bond of a saturated carbon is equally distant at an angle of 109.5°, from the other. In three dimensions, these equally spaced bonds create a tetrahedron.
- For normal or branched alkanes, C_nH_{2n+2} gives you an easy formula to determine the number of hydrogens that will bond to a particular number of carbons. For example, if you have 13 carbons, C_{13}, in a compound, you will have 2(13) + 2 hydrogens, $H_{2(13)+2}$, bonded to those carbons, $C_{13}H_{28}$.
- The **IUPAC system** is an internationally agreed system by which organic compounds are named.
- The name of a saturated hydrocarbon or alkane ends in **-ane** preceded by the prefix for the number of carbons in the longest straight chain. For example, an alkane with seven carbons is named heptane. Table 17.1 lists prefixes for up to ten carbon atoms.

Section 17.3

- **Alkenes** are compounds containing at least one carbon-carbon double bond.
- The three single bonds of an alkene are separated by 120° and arranged in a **trigonal planar** geometry.
- Simple alkenes are named by preceding the suffix -ene with the name of the longest carbon chain. C_7H_{14} is heptene.

```
    H   H   H       H   H   H
    |   |   |       |   |   |
H - C - C - C = C - C - C - C - H    3-heptene
    |   |       |   |   |   |
    H   H       H   H   H   H
```

one isomer of heptene

```
    H       H   H   H   H   H
    |       |   |   |   |   |
H - C = C - C - C - C - C - C - H    1-heptene
        |   |   |   |   |   |
        H   H   H   H   H   H
```

a different heptene isomer (C_7H_{14})

- **Dienes** are compounds containing more than one carbon-carbon double bond. Dienes are named by adding **-diene** to the proper longest carbon chain prefix. Thus C_7H_{12} is named heptadiene.
- Compounds with three or four double bonds are called **trienes** or **tetraenes,** respectively. These compounds are named similarly to the dienes.
- **Alkynes** are compounds containing at least one carbon-carbon triple bond.
- The single bonds of the triple-bonded carbon are separated by 180° forming a linear molecule.
- To name alkynes, add **-yne** to the proper prefix.
- Any compound containing a double or triple bond is said to be **unsaturated**. Without double and triple bonds more hydrogens could bond to the carbon chain thus the chain is not saturated.

Section 17.4

- A saturated ring has the formula C_nH_{2n}. Saturated rings contain no double or triple bonds.
- The optimum bond angle for saturated carbons is 109.5°. Those ring structures creating angles close to this value are the most stable. Thus cyclopropane, *cyclo* for ring, with an angle of 60° would strain the bond below the optimum value creating an unstable molecule. On the other hand, cyclopentane creates angles of 108° which provides a stable ring environment.
- The unsaturated ring compound, benzene, is particularly stable because its internal angles are 120° and its electrons experience a **de-localization** phenomenon. The electrons can freely roam the entire ring structure.
- **Condensed ring** structures share a common side such as naphthalene, $C_{10}H_{18}$.
- **Aromatic compounds** are ring structures containing alternating double and single bonds.

Section 17.5

- **Structural isomers** are compounds with the same chemical formula but differing atomic arrangements.
- The characteristics of the different structural isomers may be similar or drastically different.

Section 17.6

- **Functional groups** are singular atoms such as oxygen, nitrogen or groups of atoms such as -OH that replace hydrogen on the hydrocarbon backbone.
- These functional groups contribute immensely to the chemistry of a compound. Table 17.3 should be memorized.

Section 17.7

- An **alkyl group** is an alkane attached to another group of atoms. Alkyl groups are named by adding *-yl* to the end of the prefix noting the number of carbons in the chain.
- Alkyl groups do not always have to be straight-chained. Table 17.4 displays some branched alkyl groups with which you should be familiar.
- The IUPAC Naming System: Naming compounds.
  1. Count the longest chain of carbons including those within a functional group. Using Table 17.1, determine the appropriate name for the particular number of carbons. Look to see if the longest chain contains any double or triple bonds between carbons. If no, the base for the compound's name will end in *-ane*. If there is a double or triple bond, the base of the compound's name will end in -*ene* or *yne* respectively. Consider the following example:

$$\begin{array}{c} CH_3 \\ | \\ HO-CH_2-CH_2-CH=C-CH=CH_2 \end{array}$$

  or with all the hydrogen bonds drawn in,

```
        H   H   H   CH3     H
        |   |   |   |       |
   HO - C - C - C = C - C = C
        |   |           |   |
        H   H           H   H
```

where the methyl group is shown in abbreviated form.

```
                            H
                            |
( - CH3 )   =   (   H  -  C  -   )
                            |
                            H
```

The longest chain contains 6 carbons thus we use the name hex-. There are also double bonds, thus the compound is an alkene with the name hexene.

2. Using Table 17.5, look for the functional groups with the highest priority. Notice that an alcohol functional group, -OH, has higher priority than an alkene, -C=C-. Therefore we name the compound with *-ol* at the end. Hexene- becomes hexenol.
3. Assign numbers to the carbons in the main chain making sure to assign the lowest number possible to the highest priority functional group.

```
        H   H   H   CH3 H   H
        |   |   |   |   |   |
   HO - C - C - C = C - C = C    INCORRECT!
        |6  |5  4   3   2   |1
        H   H               H
```

Note that this is an **incorrect** number assignment. The carbon with the hydroxy functional group has priority over the double bonded carbon at the end of the chain. The following has the correct number assignment.

```
        H   H   H   CH3     H
        |   |   |   |       |
   HO - C - C - C = C - C = C    CORRECT
        |1  |2  3   4   |5  |6
        H   H           H   H
```

4. Identify groups not included in the main chain by using their prefixes in the name. We would include the prefix *methyl-* in our example.

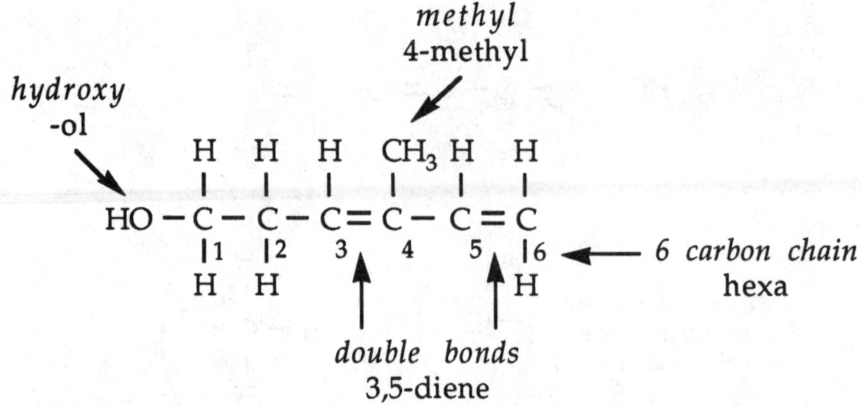

5. Use the numbers assigned to the carbons in the chain to write the complete name designating the positions of the functional groups. The compound's name ends in -ol because the hydroxy group has the highest priority of all the functional groups. We also must note in the compound name the number of double bonds. Diene, triene respectively identify two and three double bonds. Moreover, we must include the number assignment locations in the name. Hence, diene becomes 3,5-diene which tells us that the two double bonds are located at the third and fifth carbon in the backbone. The complete compound name is as follows,

4-methylhexa-3,5-dienol

This compound can also be called 4-methyl-3,5-hexadienol.

- Naming organic compounds becomes considerably easier if you memorize the names, prefixes, suffixes, and priorities for the functional groups.
- Memorizing functional group Tables also facilitates drawing structures from names.
- IUPAC Naming System: Drawing compounds.

1. Draw a backbone of the appropriate number of carbon atoms. Assign numbers to each of the carbon atoms. Direction of numbering does not matter. For example note that 4-methyl-3-bromopentanoic acid has 5 carbons (pent-). Which has a backbone as follows:

$$\underset{1}{C} - \underset{2}{C} - \underset{3}{C} - \underset{4}{C} - \underset{5}{C}$$

2. Add functional groups to the appropriate carbon as indicated by the numbers preceding the functional group prefixes and the compound name ending. Draw in the methyl group connected to carbon 4 and a bromide group connected to carbon 3.

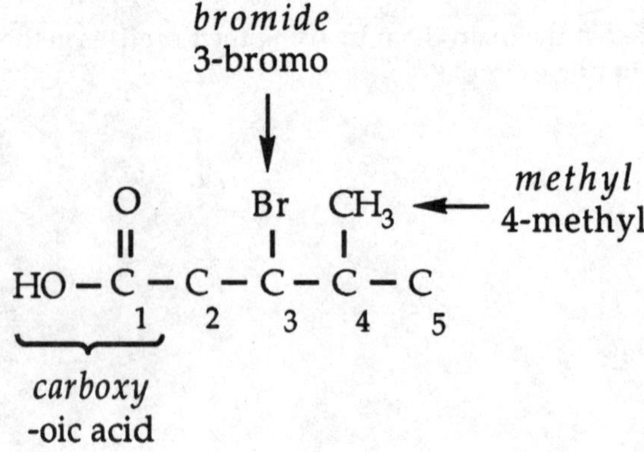

Note that we do not add another carboxylic group to the #1 carbon but make the #1 carbon into a carboxy functional group.

3. Add enough hydrogens to each carbon so that it has four bonds.

```
      O   H  Br  CH3  H
      ||  |   |   |   |
HO — C — C — C — C — C — H
          |   |   |   |
          H   H   H   H
```

4-methyl-3-bromopentanoic acid

Section 17.8

- Alkanes are not very reactive because carbon-carbon single bonds and carbon-hydrogen bonds are strong and not especially polar.
- Since the alkane carbon is bonded on all sides it is less vulnerable to attack.
- R is a short hand way of writing "the rest of the molecule." For example an alcohol can be drawn as $R\text{-}CH_2OH$ instead of always drawing out $H_3C\text{-}CH_2\text{-}CH_2\text{-}CH_2\text{-}CH_2\text{-}CH_2\text{-}CH_2OH$. This notation is handy if you need to draw out a lengthy molecule many times.
- **Substitution reactions** are reactions that replace functional groups in a molecule with a different group.
- **Elimination reactions** are reactions that usually create double bonds by losing two or more groups from a molecule.
- Both of these reactions require breakage of covalent bonds and formation of new ones.

Section 17.9

- Bonds are simultaneously broken and created in substitution reactions.

Section 17.10

- Unsaturated double bonds have a second shared electron pair that is susceptible to attacks by other atoms attracted to electron pairs.
- Double bonds undergo characteristic **addition reactions** resulting in a single bond with two atoms, or groups, newly added to the molecule.
- **Hydrogenation** is a type of addition reaction involving H_2. Unsaturated crude oil are made into useful products by this process.
- Unsaturated compounds readily add HBr and HCl creating **alkyl halides**, alkanes containing halides such as Br and Cl.
- **Markovnikov's Rule** states that when an acid adds to a double bond, the hydrogen of the acid will add to the double bonded carbon with the most hydrogens.
- **R·** is the notation for a **free radical** which has an unshared electron (one unshared electron only, not an unshared electron pair). These free radicals are especially reactive with other free radicals and double bonds. The free radical separates the susceptible electron pair of a double bond, distributing one electron to one carbon and one to the other. The free radical becomes bonded to one of the carbons still leaving the other carbon with an unshared electron. Thus the new structure is still a free radical with high reactivity. The free radicals continue reacting until all the unsaturated compounds are consumed or two free radicals happen to meet.
- **Polymerizations** are reactions, such the one described above, that repeat over and over producing large molecules, **polymers** from single units, **monomers**.
- Free radical polymerizations are responsible for the formation of Vinyl polymers such as polyethylene, polyvinylchloride (PVC), polystyrene and teflon. Thank goodness for free radicals in action.

Section 17.11

- Because of electron de-localization, aromatic benzene rings do not react as do compounds with double bonds. When benzene does react, it undergoes substitution reactions not addition reactions. For example, Br adds to a alkene whereas it replaces a hydrogen on the benzene ring in a substitution reaction. This substitution reaction requires a catalyst, $FeCl_3$, for it to react to any extent.
- Di-substituted benzene rings, two functional groups attached, are named with two different methods, using *ortho, meta, para* prefixes (abbreviated by *o*-, *m*-, and *p*-) or using numbers. If the two groups or atoms are next to each other we use ortho or 1,2. If the two groups are separated by one carbon, we use meta or 1,3. Lastly, para or 1,4 is used for the groups that are on opposite corners of the benzene ring. Note that the following compounds have the same chemical formula but different structures thus they are isomers of each other.

Br Br Br Br Br Br

1,2 (*ortho, o-*) 1,3 (*meta, m-*) 1,4 (*para, p-*)

- Some monosubstituted benzenes have common names that are used in the IUPAC naming system:

CH_3 HO

Toluene Phenol

- When one of the two groups produces a compound that has a common name, we use that name as the parent:

Br OH Cl CH_3

o-Bromophenol *p*-Chlorotoluene

- When numbering the benzene ring, do not forget to assign the lowest number to the functional group with the highest priority.
- If more than two hydrogens are replaced on the benzene ring, the numbering system is used.

Section 17.12

- **Alcohols** are hydrocarbons (except in aromatics) that have a hydrogen replaced by an -OH.
- The smaller the hydrocarbon chain the more the -OH group influences the chemical properties. On the other hand, larger alcohols behave more like their parent hydrocarbon.

name	structure	polarity	solubility with H_2O
methanol	$HO\text{-}CH_3$	polar	mixes with H_2O
1-hexanol	$HO\text{-}CH_2\text{-}CH_2\text{-}CH_2\text{-}CH_2\text{-}CH_2\text{-}CH_3$	non-polar	slightly soluble
1-octanol	$HO\text{-}CH_2\text{-}CH_2\text{-}CH_2\text{-}CH_2\text{-}CH_2\text{-}CH_2\text{-}CH_2\text{-}CH_3$	non-polar	insoluble

- Methanol, ethanol, and 2-propanol are the most commonly used alcohols. Methanol or ethanol can be used with gasoline or alone as fuel. 2-propanol is common rubbing alcohol.
- **Primary alcohols** have their -OH group on the end of the hydrocarbon. Primary alcohols can be oxidized into aldehydes. Our livers for instance, oxidize ethanol first into acetylaldehyde. Aldehydes can be further oxidized into carboxylic acids.
- **Secondary alcohols** have their -OH group on a carbon that is bonded to two other carbons. Secondary alcohols can be oxidized into ketones. Unlike aldehydes, ketones are resistant to further oxidation into carboxylic acids.

Section 17.13

- Aldehydes and ketones are similar in that they both possess a **carbonyl group**, carbon atom double bonded to an oxygen.
- The smallest and industrially most important aldehyde is methanal or commonly called formaldehyde. Though methanal is a gas at room temperature it is soluble in water. Methanal is commonly sold as a 40% solution known as formalin. Formalin kills bacteria and is used to sterilize surgical instruments. It is also used in the embalming of cadavers. More importantly, formaldehyde is widely used as a polymerization cross linker with other compounds to make synthetic plastics, adhesives, coatings and even pigment for fluorescent colors.

Section 17.14

- Amines are derivatives of ammonia where the hydrogens have been replaced by alkyl groups.
- A **primary** amine has only one carbon bonded to the nitrogen. Remember the "R-" notation is short hand for "the rest of the molecule" not important to the equation or reaction of concern.
- **Secondary** amines have two and **tertiary** amine three carbons attached to the central nitrogen.
- An amine maintains an unshared electron pair that is susceptible to attack by an electron pair acceptor or acid. Reactions between amines and acids are similar to reactions between ammonia and acids. These reactions create salts.

Section 17.15

- Carboxylic acids are important acidic compounds and participate in polymerization reactions.
- Stearic and lauric acids, both carboxylic acids, can be found in a common household item, shampoo.
- Carboxylic acids undergo reactions with alcohols eliminating a H_2O while creating an ester.
- Esters have an odiferous nature and lend their pleasant-smelling nature to bubble gum, fruits, flowers and perfumes.

Section 17.16

- A water molecule is also eliminated when a carboxylic acid reacts with an amine producing a larger molecule called an amide.
- Not only is this type of reaction important in industrial production of synthetic fibers, such as dacron, it also provides the link between monomers of proteins. Proteins will be discussed in the next chapter.
- In a **condensation reaction** two monomers, such as a carboxylic acid and an alcohol or amine, combine to form a larger molecule, a condensation polymer. A condensation reaction may also eliminate a small molecule other than H_2O. For example, the condensation reaction producing a type of nylon releases hydrogen chloride.
- A monomer with only one functional group allows only one link to form thus it produces a dead end. However, if the monomer has more than one reactive functional group, then more connections can be made. Only one condensation reaction can occur with two difunctional monomers whereas **polyfunctional** molecules are capable of creating longer and more complex networking polymers.

Section 17.17

- The fuel you pump into your car is a homogeneous mixture containing hydrocarbons of chain lengths 6, 9, or 10 carbons. These chain lengths are separated from the crude petroleum through a process called refining. The different chain lengths influence the hydrocarbon's boiling point thus they can be separated from each other by heating the crude.

B. Solutions to In-Text Exercises

Exercise 17.1

Table 17.1 gives us the prefix for a six carbon chain; hex- = 6 carbons. The exercise already notes that the compound is an alkene which have endings of *-ene*. Thus the alkene C_6H_{12} is hexene.

Exercise 17.2

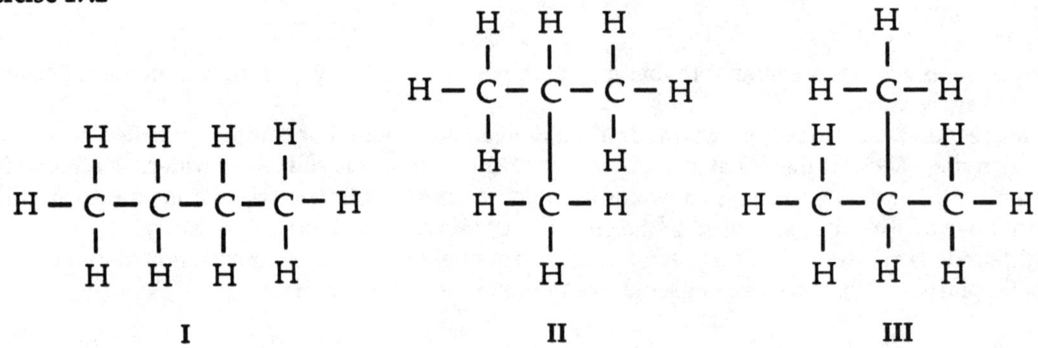

Structures II and III are the same structural isomer rotated 180°.

Exercise 17.3

The question only requires 5 different structural isomers.

1.

```
    H   H   H   H   H             H           H   H
    |   |   |   |   |             |           |   |
H - C - C = C - C - C - H     H - C - C = C - C - C - H
    |           |   |             |   |   |   |   |
    H           H   H    and      H   H   H   H   H
```

are the same.

2.

```
    H   H   H   H   H             H   H   H   H   H
    |   |   |   |   |             |   |   |   |   |
H - C = C - C - C - C - H     H - C - C - C - C = C - H
            |   |   |             |   |   |
            H   H   H    and      H   H   H
```

are the same.

3.

```
    H       H   H
    |       |   |
H - C - C = C - C - H
    |   |       |
    H   |       H
        |
    H - C - H
        |
        H
```

4.

```
    H       H   H              H   H       H
    |       |   |              |   |       |
H - C = C - C - C - H      H - C - C - C = C - H
        |   |   |              |   |   |
        |   H   H              H   H   |
        |                              |
    H - C - H                      H - C - H
        |                              |
        H              and             H
```

are also the same.

5.

```
    H   H   H
    |   |   |
H - C - C - C = C - H
    |   |       |
    H   |       H
        |
    H - C - H
        |
        H
```

6.

```
        H     H
   H     \   /     H
    \      C      /
H -  C   /   \   C - H
      \         /
   H - C ----- C - H
       |       |
       H       H
```

Exercise 17.4

1. The longest chain consists of three carbons. Find the appropriate carbon chain name from Table 17.1; pro- = 3 carbons.

2. Note the functional group for which the compound will be named. We note that the aldehyde has priority over bromine from Table 17.5. Thus the compound will have an ending *-al*.

```
                 aldehyde
   H   H         higher priority
   |   |   ______
H— C — C — C=O
   |   |   |
   Br  H   H
   ↑
bromine
```

3. The carbon with the aldehyde group is assigned the lowest possible number. This assignment gives the carbon with bromide attached a number 3.

```
   H   H
   |   |
H— C — C — C=O
   |1  |2  |3
   Br  H   H
```

This number assignment gives bromine the higher priority which is incorrect. The following is the correct assignment.

```
   H   H
   |   |
H— C — C — C=O
   |3  |2  |1
   Br  H   H
```

4. Now we have all the information necessary to write the full name. The ending propanal results from the pro- = 3 carbon chain, -pane = alkane (no double or triple carbon bonds), and -al for the aldehyde.

```
                 aldehyde
   H   H         -al
   |   |   ______
H— C — C — C=O
   |3  |2  |1
   Br  H   H
   ↑
bromine
3-bromo-
```

3-bromopropanal

Exercise 17.5

Step 1. The carbon ring has 5 carbons. Use the prefix cyclo- to indicate the ring form and penta- for the 5 carbons.

Step 2. The only functional group is the ketone thus the name will end with -one.

Step 3 and 4 not needed in this exercise.

Step 5. Put all the prefixes and suffixes together.

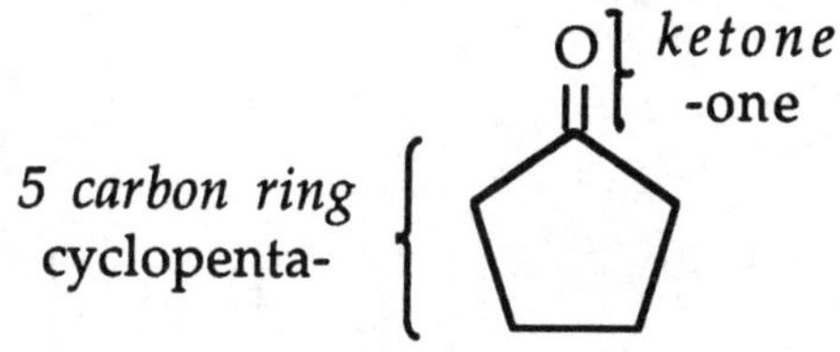

cyclopentanone

Exercise 17.6

Step 1. The longest chain has 4 carbons; buta = 4 carbons.

Step 2. The functional group with higher priority is the carboxylic acid thus the name will end with -oic acid.

Step 3. Start at the carboxylic carbon and assign numbers to the carbons.

Step 4. The bromine group is at the third carbon thus we use 3-bromo.

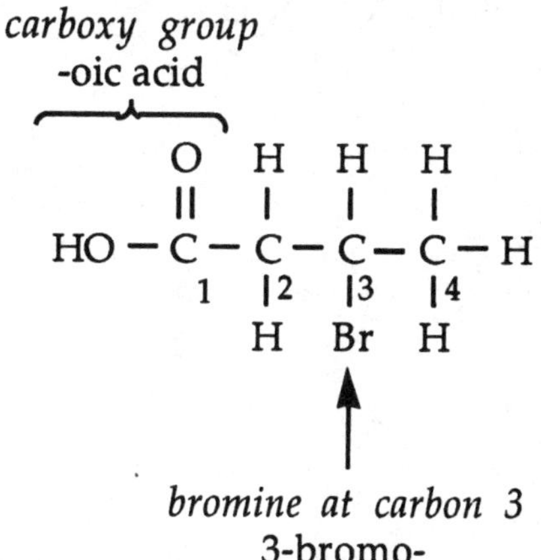

Step 5. 3-bromobutanoic acid. We use the *n* in noic to make the name easier to pronounce.

Exercise 17.7

Step 1. -octa means there are 8 carbons in the main chain.

```
C – C – C – C – C – C – C – C
1   2   3   4   5   6   7   8
```

Step 2. Add in the functional groups at the appropriate carbon positions.

Step 3. Fill in the hydrogens.

```
                          H
                          |
                      H – C – H    } 5-ethyl-
                          |
                      H – C – H
-oic acid  O   H   H   H  |                 H
           ||  |   |   |  |                 |
      HO – C – C – C – C – C – C = C – C – H
           1   |2  |3  |4  |5  |6   |7  |8   ↖
               H   H   |   H   H  ↖H    H      -oct-
                   H – C – H        6-......en-
                       |
       -4-methyl       H
```

Exercise 17.8

Step 1. Cyclohepta- means there are 7 carbons in a ring.

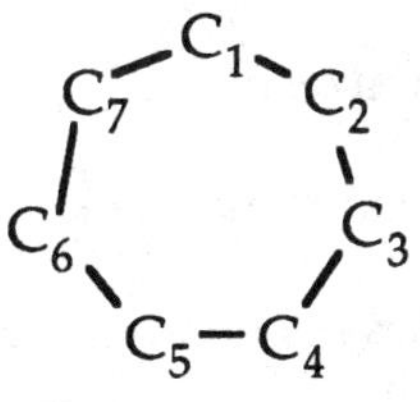

C_1 C_2 C_3 C_4 C_5 C_6 C_7

Step 2. The ending -one means the compound is named for a ketone. This ketone has priority over all the methyl groups thus the ketone will be found on the number one carbon. 2,4,6-trimethyl- means there are 3 methyl groups; one at carbon 2, 4 and 6.

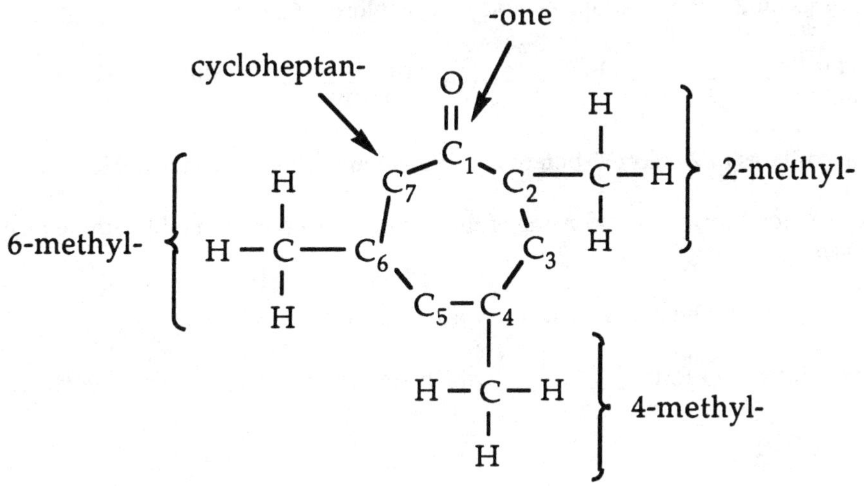

-one

cycloheptan-

2-methyl-

6-methyl-

4-methyl-

Step 3. Fill in the hydrogens.

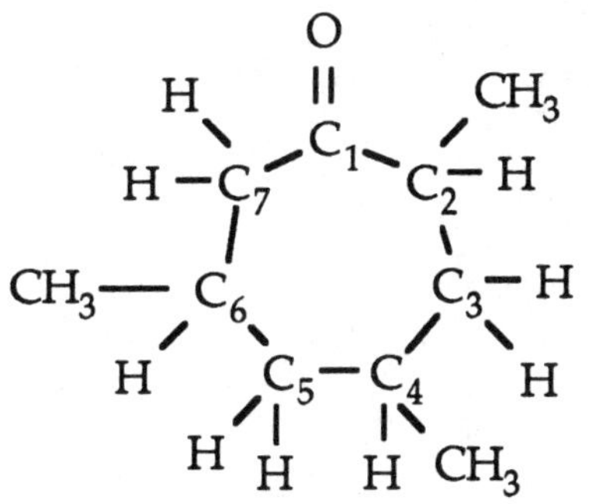

C. Chapter 17 Practice Test

1. The general formulas for alkanes and alkenes are _________ and _________ , respectively.

2. Draw all structural isomers of $C_3Cl_3H_5$ and name them.

3. Write structural formulas for each of these compounds: 2-chloro-3-methylbutan-2-ol, 1-bromo-2-chlorohexane, and pentanoic acid.

4. Write the structural formula and name for the product obtained when 1-butene reacts with hydrogen bromide.

5. Draw structures for 2,4-dichlorotoluene and *para*-dichlorobenzene.

6. Alcohols can be ___________ to ___________ or _________ using dichromate or permanganate.

7. Construct a synthesis of 2-chloro-2-butanol from a ketone. Name the starting ketone.

8. Predict the structural formulas and name of the product formed when ethyl amine reacts with propanoic acid.

9. An _________ forms when an acid reacts with an alcohol.

10. Monomers condense to form _________ , which are high molecular weight molecules.

D. Answers to Practice Test

1. C_nH_{2n+2}, C_nH_{2n}

2.

$$\begin{array}{ccccccc} & & H & & H & & Cl \\ & & | & & | & & | \\ Cl & - & C & - & C & - & C - Cl, \\ & & | & & | & & | \\ & & H & & H & & H \end{array}$$

1,1,3-trichloropropane

$$\begin{array}{ccccccc} & & H & & H & & Cl \\ & & | & & | & & | \\ H & - & C & - & C & - & C - Cl, \\ & & | & & | & & | \\ & & H & & Cl & & H \end{array}$$

1,1,2-trichloropropane

$$\begin{array}{ccccccc} & & Cl & & Cl & & Cl \\ & & | & & | & & | \\ H & - & C & - & C & - & C - H, \\ & & | & & | & & | \\ & & H & & H & & H \end{array}$$

1,2,3-trichloropropane

$$\begin{array}{ccccccc} & & H & & H & & Cl \\ & & | & & | & & | \\ H & - & C & - & C & - & C - Cl, \\ & & | & & | & & | \\ & & H & & H & & Cl \end{array}$$

1,1,1-trichloropropane

$$\begin{array}{ccccccc} & & H & & Cl & & H \\ & & | & & | & & | \\ Cl & - & C & - & C & - & C - H \\ & & | & & | & & | \\ & & H & & Cl & & H \end{array}$$

1,2,2-trichloropropane

3.

$$\begin{array}{ccccccccc} & & H & & Cl & & CH_3 & & H \\ & & | & & | & & | & & | \\ H & - & C & - & C & - & C & - & C - H \\ & & | & & | & & | & & | \\ & & H & & OH & & H & & H \end{array}$$

2-chloro-3-methylbutan-2-ol

$$\begin{array}{ccccccccccccc} & & H & & Cl & & H & & H & & H & & H \\ & & | & & | & & | & & | & & | & & | \\ Br & - & C & - & C & - & C & - & C & - & C & - & C - H \\ & & | & & | & & | & & | & & | & & | \\ & & H & & H & & H & & H & & H & & H \end{array}$$

1-bromo-2-chlorohexane

$$\begin{array}{ccccccccccc} & & H & & H & & H & & H & & O \\ & & | & & | & & | & & | & & \| \\ H & - & C & - & C & - & C & - & C & - & C - OH \\ & & | & & | & & | & & | & & | \\ & & H & & H & & H & & H & & H \end{array}$$

pentanoic acid

4.

```
      Br  H   H   H                 H   Br  H   H
      |   |   |   |                 |   |   |   |
  H - C - C - C - C - H   or   H -  C - C - C - C - H
      |   |   |   |                 |   |   |   |
      H   H   H   H                 H   H   H   H
```

1-bromobutane — 2-bromobutane

(2-bromobutane is the product formed in greatest amount because of Markovnikov's rule.)

5.

CH_3

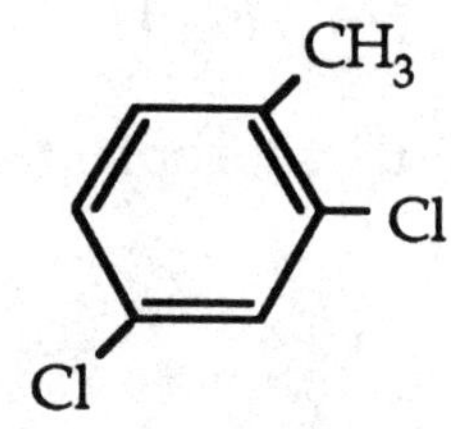

Cl

Cl

2,4-dichlorotoluene

Cl

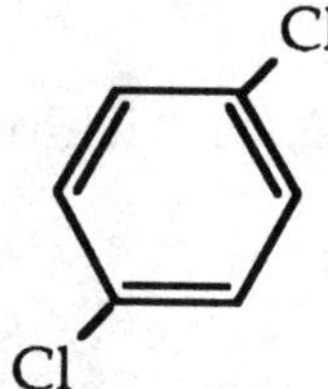

Cl

para-dichlorobenzene

6. oxidized, aldehydes, carboxylic acids

7.

```
      H   H   O   H                                  H   H   OH  H
      |   |   ||  |       Ni, high pressure          |   |   |   |
  H - C - C - C - C - H  ------------------->   H -  C - C - C - C - H
      |   |       |              HCl                 |   |   |   |
      H   H       H                                  H   H   Cl  H
```

2-butanone — 2-chloro-2-butanol

8.

```
      H   H   O       H   H
      |   |   ||      |   |
  H - C - C - C - N - C - C - H
      |   |       |   |   |
      H   H       H   H   H
```

diethylamide

9. ester

10. polymers

Chapter 18

The Chemistry of Living Organisms

A. Outline and Study Hints

Section 18.1

- The fundamental building units of living organisms are the **cells.** The cells of the body are analogous to the bricks that make up a brick house.
- The cells of living organisms are composed of common components. The following is a diagram of an animal cell. Underneath their labels, write in the appropriate function of that particular cell component.

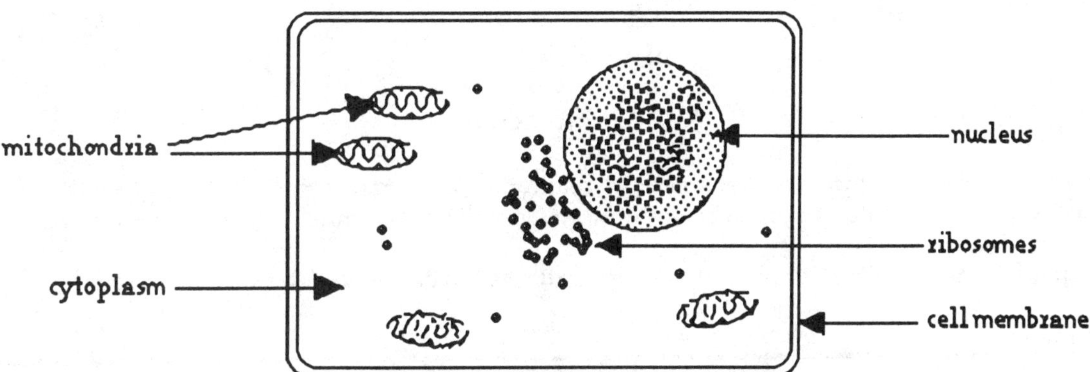

Section 18.2

- Chlorophyll is the solar power house of plant cells. Note that chlorophyll is a substance not a cell component. Chlorophyll can utilize the energy in sunlight to convert H_2O and CO_2 into glucose, a useful fuel for organisms. This amazing process is called **photosynthesis**. Animal cells cannot make their own energy as plant cells do. Animal cells must gain their energy by reversing the photosynthetic reaction, breaking down glucose into H_2O, CO_2 and energy. This break down process is called respiration.

- Photosynthesis:

$$6\ H_2O + 6\ CO_2 + \text{light energy} \rightarrow C_6H_{12}O_6 + 6\ O_2$$

- Respiration:

$$6\ H_2O + 6\ CO_2 + \text{energy} \leftarrow C_6H_{12}O_6 + 6\ O_2$$

or properly written,

$$C_6H_{12}O_6 + 6\ O_2 \rightarrow 6\ H_2O + 6\ CO_2 + \text{energy}$$

- The body does not literally burn glucose. Glucose is broken down into energy through a series of complex chemical reactions which in total are known as the **Kreb cycle** or the **citric acid cycle**. Kreb was awarded the 1953 Nobel price for elucidating this vital cycle.
- **ATP** is the carrier for the chemical energy produced by respiration. When this energy is used, a phosphate ion is released producing **ADP**, ATP with one less phosphate.

Section 18.3

- Remember from section 17.15, a condensation reaction occurs between a carboxylic acid and an amine producing an amide.
- Notice that an amino acid has both a carboxy and amino functional group thus they are called amino (for the amine) acids (for the carboxylic acid).

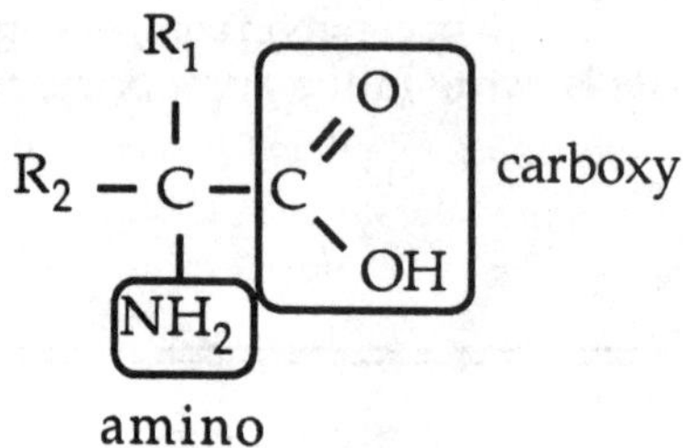

Amino acid

- The amide produced during condensation polymerization between two amino acids is given the special name of **peptide**. The bond between the amino and carboxy group is called a **peptide bond** or peptide linkage.
- Proteins are polypeptides, or many amino acids attached to one another by peptide linkages.

Section 18.4

Proteins			
NON-STRUCTURAL	Fibrous	keratins	hair skin feather claws wool
	Muscular	collagens	tendons connective tissue such as cartilage
GLOBULAR	Enzymes		Highly specific catalysts, essential to the chemistry of living organisms
	Respiratory proteins		Used for O_2 and CO_2 transport and storage; an example is hemoglobin
	Nucleoproteins		Function together with DNA and RNA to store and transmit the information of heredity
	Antibodies		Protect the organism from harmful substances such as viruses and the byproducts of bacterial metabolism
	Hormones		Important regulatory proteins that act as controls for the chemical process of the body

- Proteins have structure that is categorized. For example, the primary structure of a protein is just its sequence of amino acids whereas the quaternary level is an arrangement of separate protein units. The secondary structure is a particular folding pattern while the tertiary structure is additional folding of the secondary structure into a three dimensional structure.

Section 18.5

- **DNA, deoxyribonucleic acid,** could arguably be the most important polymer in the body without which the human organism could not accomplish its functions. Without DNA, the cell lacks the appropriate information to produce the proteins needed for the cell's functions. Memorize and understand all you can about this beautifully significant polymer and its co-workers, RNA and ribosomes, to mention a couple.
- There are four primary bases (adenine, A; guanine, G; thymine, T; cytosine, C) in DNA. There are also four bases in RNA. As in DNA, adenine, guanine, and cytosine are found in RNA. However, uracil replaces thymine in RNA.
- The sugar component of DNA is deoxyribose whereas RNA contains ribose.
- DNA and RNA are both polymers of **nucleotide** monomers; nucleo-for nucleic acid and -tide for sugar. Each monomer of RNA and DNA is composed of a phosphate group, a sugar molecule, and a base group.
- The tertiary structure of DNA is in the form of a **double helix**. Two strands of DNA are spiraled around each other such that their bases are arrange toward the central axis of the helix. RNA usually is single stranded and has different tertiary structures depending on the type.
- The processes of **replication, transcription,** and **translation** are shown in the following schematic.

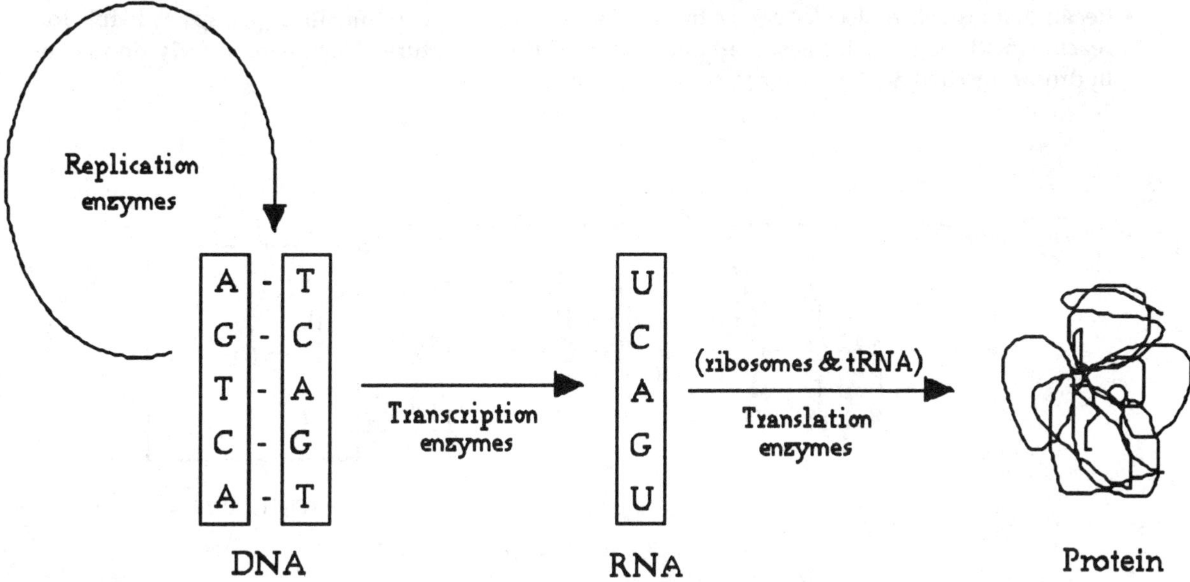

- Note DNA can be duplicated by **replication** or "read" to produce RNA via **transcription**. This RNA strand, which has the same sequence (U replacing T) as one of the DNA strands, is used in **translation** of the base pair sequence to proteins.

Section 18.6

- From the last section, you might have wondered how does DNA get replicated and RNA translated. DNA and RNA cannot achieve these processes on their own. They require a collage of proteins that catalyze individual reactions in the complete process of replication and translation. These proteins are called **enzymes**. Enzymes are proteins that catalyze a particular reaction or set of similar reactions. These proteins allow for very efficient and quick control of the body's functions.

- Without enzymes, replication, transcription or translation would not occur and the cell could not duplicate itself. A defective enzyme involved in these processes would also be fatal.

Section 18.7

- Hormones are compounds secreted by specific glands, released into the blood stream and used as messengers to alter or direct reactions in the body. For example, insulin is a hormone secreted by the pancreas. It regulates the metabolism of carbohydrates, lipids and amino acids. Insulin dependent Diabetes Mellitus patients totally lack insulin activity and must have several injections of this important hormone in order to survive.
- Besides repication and transcription, enzymes serve as catalysts for a host of other processes. For example, the enzyme alcohol dehydrogenase will catalyze the reaction of an alcohol to an aldehyde in a living system.

Section 18.8

- Recall from section 18.2 that respiration breaks down glucose into CO_2, H_2O and energy.
- Glucose is a singular sugar unit, a **monosaccharide**. A two unit sugar, sucrose, is classified as a **disaccharide** (di-for two sugars). Sucrose is made up of glucose and fructose.
- Note that the three most important carbohydrates are polymers of glucose known as polysaccharides-starch, cellulose and glycogen. In digestion, starch and glycogen glucose branches are broken into smaller and smaller polysaccharides and eventually into monosaccharides of glucose. The human body lacks the enzyme that can recognize the particular branching pattern of cellulose, thus our bodies cannot use cellulose as a glucose energy source. Most of the glucose we use as energy comes from the digestion of carbohydrates.
- Recall that organic molecules can be bi-functional possessing two functional groups available for reaction with other molecules. Fatty acids have this dual nature. They have a **fatty** (long chain hydrocarbon) and **acidic** (carboxylic region or head) region.

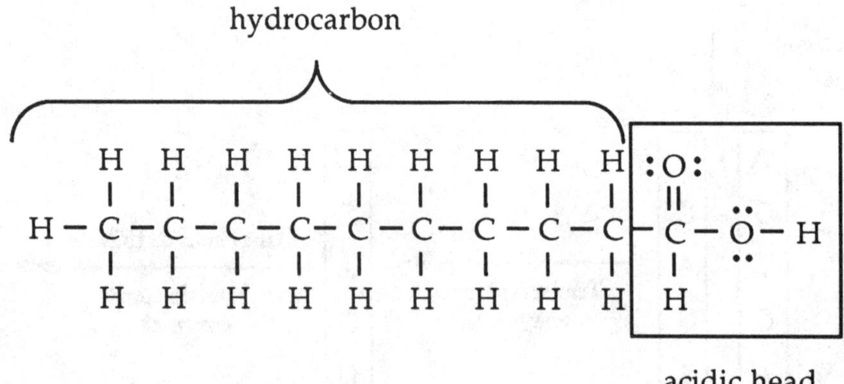

Lauric acid

- Note that the acidic portion is very hydrophilic (loves water) while the hydrocarbon tail is hydrophobic (repels water). Thus water will be attracted to the acidic head while nonpolar molecules such as oil or dirt will be attracted to the hydrophobic portion of the fatty acid. This is how detergents work. Dirt particles get whisked away in the soapy water because dirt particles get surrounded and trapped inside a sphere of fatty acids. The surface of the sphere consists of protruding acidic head groups while the inner portion of the sphere contains all the hydrocarbon tails. It might appear strange that the body needs molecules with "soapy" properties. On the contrary, complex fatty acids with their polar heads and non-polar bodies are essential components of the cell membrane.

B. Solutions to In-Text Exercises

There are no In-Text exercises for this chapter.

C. Chapter 18 Practice Test

1. The ______ ______ acting as a barrier surrounds the cell. ______ fills the cell in which the ______ (powers houses), ______ (factories of cell compounds) and ______ (brains of the cell) are suspended.

2. Photosynthesis converts CO_2 and H_2O into stored energy in the form of glucose. True or false?

3. Write the balanced equation for the respiration of glucose.

4. ADP is an immediate molecular source of energy for a cell. True or false?

5. Proteins are polymers made of ____-_________ ________ .

6. What kind of polymerization occurs when a polypeptide is made? The peptide bond occurs between what two functional groups on the two amino acids?

7. What are the eight essential amino acids and why are they considered essential?

8. List two differences between DNA and RNA. What does DNA and RNA stand for?

9. How many sugar units are in monosaccharides, disaccharides, trisaccharides and polysaccharides?

10. What is the difference between saturated fats and unsaturated fats? How does this affect packing of the molecules ? Which would more likely be solid at room temperature?

D. Answers to Practice Test

1. cell membrane, cytoplasm, mitochondria, ribosomes, nucleus

2. True

3. $C_6H_{12}O_6 + 6\ O_2 \rightarrow 6\ H_2O + 6\ CO_2$ + energy

4. False; ATP is

5. α amino acids

6. condensation; carboxy and amino groups

7. val, leu, ilu, thr, phe, try, lys, met; Humans cannot synthesize these amino acids and must extract them from their diet.

8. Deoxyribonucleic acid (DNA) has one less hydroxy group. Ribonucleic acid (RNA) has uracil instead of thymine.

9. one, two, three, many

10. The long chains of saturated fats lack double carbon-carbon bonds allowing regular packing alignment and high dipole interactions. These high dipole interactions lead to high melting temperatures. Unsaturated fats have double bonds causing irregular packing and less intermolecular interaction hence lower melting points. Saturated fats would more likely be solid at room temperature.

Chapter 19

Chemistry and the Atomic Nucleus

A. Outline and Study Hints

Section 19.1

- Study Table 19.1 closely, memorizing the symbols for the different radiation particles. Also note the individual characteristics that allow the different penetration levels.
- Know the contributions to nuclear chemistry from Henri Becquerel, Marija Curie (Madame Curie) and Ernest Rutherford.
- Wilhelm Roentgen (1845-1923) discovered x-rays in 1895.

Section 19.2

- Remember the definition of an isotope. They are atoms with a different number of neutrons but with identical number of protons. Thus an atom with 6 protons and 6 neutrons and another with 6 protons and 7 neutrons are both carbon isotopes.
- Because of the identifying proton number, we can write carbon-12 or carbon-13 knowing the atomic number for both isotopes is six.

Section 19.3

- Short lived isotopes, those with short half-lives, can be quite useful in medical treatment and diagnosis. Some common medically-employed isotopes are $^{99m}_{43}Tc$, $^{32}_{15}P$, $^{131}_{53}I$ and $^{197}_{80}Hg$.

Symbol	Isotope	Half-life	Use
$^{99}_{43}Tc$	Technetium-99m	6.0 hours	Brain, liver, kidney, bone marrow scans; diagnosis of damaged heart muscles.
$^{32}_{15}P$	Phosphorus-32	14.3 days	Detection of eye tumors.
$^{131}_{53}I$	Iodine-131	8 days	Detection of thyroid malfunction; treatment of hyperthyroidism and thyroid cancer.
$^{197}_{80}Hg$	Mercury-197	65 hours	Kidney scan

Section 19.4

- Note that radiocarbon dating is only informative on once living samples. Only when the organism has died does the amount of $^{14}_{6}C$ start to decrease due to decay.
- Other types of radioactive dating are used to date samples that have never lived. For example, rocks can be dated by their lead-206 and uranium-238 content.

Section 19.5

- Memorizing all the types of emitted particles, their names, symbols, atomic number and mass number, given in Table 19.1 greatly helps you write nuclear equations correctly and quickly.
- Furthermore, do all the Section 19.5 problems at the end of the chapter and you will be efficient at writing balanced nuclear equations.
- Remember **transmutation** occurs when an element alters its atomic number, the number of protons, creating a different element.

Section 19.6

- The decay rate of a radioactive element **cannot** be decreased nor accelerated, but a proton can be accelerated to such high energy that it will transmute an element when collided.

Section 19.7

- An atom needs a huge amount of force in order to keep the positive particles so closelly packed in the nucleus. Remember that like charges repel each other.
- The atom gains the force necessary by converting mass to energy. If you weigh the parts (proton and neutron) of one ${}^{2}_{1}H$, you find that the sum of the parts does not equal the whole. The mass of one ${}^{2}_{1}H$ is actually less that the sum of its parts. Where did the mass go? It was converted into a large amount of energy according to Einstein's equation,

$$E = mc^2$$

where *E* is energy in joules, *m* si mass in kilograms, and*c* is the speed of light (3 × 10 m/s). A small amount of mass will result in a large energy value because of the large constant *c*.

- The difference in the mass of a nucleus and the combined masses of its protons and neurtons taken as seperate particles is called **mass defect**.

- **Binding energy of the nnucleus** is the mass defect converted to energy units using the following conversion factor:

$$\frac{1.5 \times 10^{-13}\ \text{kJ}}{\text{amu}} = \frac{9 \times 10^{13}\ \text{kJ}}{\text{kg}}$$

- $E = mc^2$ was messaged algebraically to get these convenient conversion factors. Note the conversion factor use in examples 19.6 and 19.7.

Section 19.8

- **Nuclear fission** is the fragmentation of large nuclei into smaller nuclei with the release of energy. The most important product of a nuclear fission reaction is the resulting energy. This energy is produced because the products have less mass than the reactants. Remember $E = mc^2$ and defect mass.
- **A chain reaction** is a nuclear fission reaction that self-propagates. The products (neutrons) of the first collision cause further collisions and disintegration that once again produce neustrons capable of starting more reactions. Uncontrolled nuclear fission chain reactions were created in the first atomic bomb. Chain reactions can be slowed by inserting voron control rods to absorb the neurtons thus removing them from further collision.

Section 19.9

- Though the most usefull product of fission is energy, hazardous isotopes may also be produced. Note the omnious sampling of dangerous fallout isotopes in Table 19.2.

Section 19.10

- **Fusion reactions** occur when two light nuclei combine to produce a larger nucleus. Fusion reactions also release large amounts of energy but they require extreme temperatures and high speeds in order to make the nuclei fuse. Super colliders are used to accellarate nuclei to high speeds needed in fusion reactions.

Section 19.11

- Fission reactions can be controlled and regulated in **nuclear reactors**. Nuclear reactors generally use enriched uranium as the fuel elements. The feul is seperated by moderators that adjust the speed of neutrons to the most efficient speed to continue the reaction. The reaction can be slowed or completely halted with insertion of metal control rods. These metal rods absorb the neutrons so they are no longer available for further fision reactions.
- The energy produced by the plant is redirected to heat water and make high-pressure steam. The steam turns turbines and generator which in turn produce energy. Today about ten percent of the electricity used in the United States is supplied by nuclear power.
- Nuclear power pros and cons:

Pro	Con
no fossil fuel	waste: nuclear waste extremely difficult to dispose
atmospheric polution: produces no gases, ashes, or smoke	production of extremely toxic materials and bomb precursors
cost: currently fissionable isotopes cost less than oil	unsafe designs
safety: high safety record	possible accident catastrophe due to acts of war or sabotage

B. Solutions to In-Text Exercises

Exercise 19.1

$\frac{65 \text{ counts per second}}{257 \text{ counts per second}} = \frac{1}{4}$ Thus there is only a quarter of the original intensity.

$\left(\frac{1}{2}\right)^{\text{\# of half-lives}} = \frac{1}{4}$; we must solve this for the number of half-lives, but

$\left(\frac{1}{2}\right)^{2} = \frac{1}{4}$,thus we know that it takes 2 half-lives to reduce the count to a quarter of its original value.

You could also multiply the original count value by $\frac{1}{2}$ until you reach the resulting count. Of course, this takes a lot longer.

$\frac{1}{2}(257) = 128.5$ (once)

$\frac{1}{2}(128.5) = 64.25$ (twice, thus two half-lives are required)

Now calculate the number of years that corresponds to two half-lives of cobalt-60.

$(2 \text{ half-lives})\left(\frac{5.3 \text{ years}}{\text{half-life}}\right) = 11 \text{ years.}$

Exercise 19.2

Step 1: Determine how much of the original sample is left which is found by the ratio

$$\frac{72 \text{ counts per hour}}{1152 \text{ counts per hour}} = \frac{1}{16}$$

Determine how many half lives it took for the decrease in counts per hour.

$\left(\frac{1}{2}\right)^{\#\text{ of half-lives}} = \frac{1}{16}$; we must solve for # of half-lives, but

$\left(\frac{1}{2}\right)^{4} = \frac{1}{16}$, thus we know that it takes 4 half-lives to decrease the count of this particular sample of $^{131}_{53}I$ from 1152 to 72 counts per hour.

Step 2: Determine the number of days 4 half lives takes.

$$4 \text{ half-lives} \times \frac{8.05 \text{ days}}{\text{half-life}} = 32.2 \text{ days.}$$

Exercise 19.3

Step 1: Determine the amount of decay that occurred.

$$\frac{23 \text{ counts per minute}}{736 \text{ counts per minute}} = \frac{1}{32}$$

Step 2: Calculate the number of half-lives that decreases the amount of emission to $\frac{1}{32}$ the original.

$\left(\frac{1}{2}\right)^{\#\text{ of half-lives}} = \frac{1}{32}$ Solve for # of half-lives.

Solving for # of half-lives with algebra would require logarithms. However we can calculate the # of half-lives be determining to what power you have to raise 2 in order to get a value of 32. With trial and error, we see that $2^3 = 8$, $2^4 = 16$ and $2^5 = 32$. Thus

$\left(\frac{1}{2}\right)^{5} = \frac{1^5}{2^5} = \frac{1}{32}$ and the # of half-lives is 5.

Finally calculate how long one half-life is.

$\frac{71.5 \text{ days}}{5 \text{ half-lives}} = 14.3$ days is the half-life of $^{32}_{15}P$.

Exercise 19.4

Step 1: Write the parent isotope.

$^{199}_{83}Bi$

Step 2: Draw an arrow from the parent isotope to the right

$^{199}_{83}Bi \rightarrow$

and write in the particle emitted.

$^{199}_{83}Bi \rightarrow {}^{4}_{2}He$

Step 3: Write in a plus sign. Write in the mass number of the parent isotope minus the mass number of the emitted particle. In this case 199 - 4 = 195. 195 is the superscript of the daughter isotope. Now do the same procedure for the atomic number; 83 - 2 = 81. This number is the subscript for the new element. Your equation should look like this:

$^{199}_{83}Bi \rightarrow {}^{4}_{2}He + {}^{195}_{81}$

Step 4: Lastly, look up the element in the periodic table that corresponds to the daughter isotope atomic number. Tl, thallium has an atomic number of 81. Write in the symbol of the element next to the superscript and subscript numbers.

$^{199}_{83}Bi \rightarrow {}^{4}_{2}He + {}^{195}_{81}Tl$

Verify that the mass numbers and atomic numbers are balanced on each side of the arrow.

Exercise 19.5

Determine the daughter isotope and atomic number. In this example they are 15 and 7 respectively.

$^{14}_{7}N + {}^{1}_{0}n \rightarrow {}^{15}_{7}$

"But wait," you say, "the product is still nitrogen."

$^{14}_{7}N + {}^{1}_{0}n \rightarrow {}^{15}_{7}N$

Your are right. We still have to produce $^{14}_{6}C$.

$^{15}_{7}N \rightarrow ? + {}^{14}_{6}C$

Notice that $^{14}_{6}C$ has one less atomic and mass number than $^{15}_{7}N$. Remember a proton has one atomic number and one mass number. Thus we could balance our equation by adding $^{1}_{1}H$ into the product side.

$^{15}_{7}N \rightarrow {}^{1}_{1}H + {}^{14}_{6}C$

The complete equation is

$^{14}_{7}N + {}^{1}_{0}n \rightarrow {}^{1}_{1}H + {}^{14}_{6}C$ omitting the transitory $^{15}_{7}N$.

Exercise 19.6

Note the atomic number has to equal 43 on both sides of the equation. The mass number for the daughter isotope must equal 97. Write in these numbers as subscripts and superscripts, respectively.

$$^{96}_{42}Mo + ^{2}_{1}H \rightarrow ^{1}_{0}n + ^{97}_{43}$$

Look up the element that corresponds to atomic number 43 which is Tc, technetium and write in the symbol next to the atomic and mass numbers.

$$^{96}_{42}Mo + ^{2}_{1}H \rightarrow ^{1}_{0}n + ^{97}_{43}Tc$$

This equation demonstrates that bombarding molybdemum (Mo) with deuterium ($^{2}_{1}H$) produces technetium (Tc) while releasing a neutron.

Exercise 19.7

Step 1: Find the mass defect per nucleus.

$$^{7}_{3}Li + ^{1}_{1}H \rightarrow 2\,^{4}_{2}He$$

7.01601 amu + 1.00797 amu $\rightarrow$ 2(4.00260 amu)

8.0240 amu $\rightarrow$ 8.0052 amu

8.0240 amu -8.0052 amu = 0.01878 amu lost as energy.

Step 2: Convert the change in mass defect to energy.

(0.01878 amu) $\times$ (1.5 $\times$ 10^{-13} kJ/amu) = 2.8 $\times$ 10^{-15} kJ for every 2 $^{4}_{2}He$ nuclei.

$$2\text{ mol }^{4}_{2}He\left(\frac{2.8\times10^{-15}\text{ kJ}}{2\,^{4}_{2}He\text{ nuclei}}\times\frac{6.023\times10^{23}\,^{4}_{2}He\text{ nuclei}}{\text{mol }^{4}_{2}He}\right) = 1.7\times10^{9}\text{ kJ}$$

{ enough energy to melt 877 metal automobiles

Exercise 19.8

240 - 135 - 2 =103, the mass number and 94 - 51 = 43, the atomic number of the daughter isotope.

$$^{240}_{94}Pu \rightarrow ^{135}_{51}Sb + ^{103}_{43}Tc + 2\,^{1}_{0}n$$

Exercise 19.9

$$2\,^{3}_{2}He \rightarrow ^{4}_{2}He + 2\,^{1}_{1}H$$

C. Chapter 19 Practice Test

1. Why are gamma rays not affected by a magnetic field?

2. Gamma rays are not ______ of matter, but are streams of extremely ______ ______.

Match the following numbers with the appropriate letters.

3. ____ nuclear fusion

4. ____ critical mass

5. ____ beta particle

6. ____ transmutation

7. ____ nuclear fission

8. ____ alpha particle

a. a change of a radioactive element into a different element.

b. $^{0}_{-1}e$

c. the disintegration of large nuclei to form smaller nuclei with release of energy

d. $^{4}_{2}He$

e. the combination of two light nuclei to produce a larger nucleus

f. enough neutrons are present to start a chain reaction culminating in an explosion

9. Iodine 131, $^{131}_{53}I$ has a half-life of 8 days. A 10 gram amount of iodine-131 was injected into a patient for thyroid imaging. How many days would it take for the iodine to decay to a total amount of 0.3 g iodine-131?

10. Radioactive C-14 dating showed that the flax from which the Turin linen was made was living sometime between 1260 AD and 1380 AD. C-14 has a half-life of 5730 years.

   a. How many $^{14}_{6}C$ half-lives have occurred between 1993 and 1260 AD?

   b. How many particles per minute would be emitted from a 1 g pure carbon sample taken from the Turin today?

   c. How many particles per minute would have been emitted from the same sample if it was 1993 years old?

D. Answers to Practice Test

1. Gamma rays are not charged.

2. particles, high-energy photons

3. e

4. f

5. b

6. a

7. c

8. d

9. 40 days

10. 0.128 half-lives, 14.6 particles/minute, 12.6 particles/minute